Soil Physical Environment and Plant Growth

Pradeep K Sharma • Sandeep Kumar

Soil Physical Environment and Plant Growth

Evaluation and Management

Springer

Pradeep K Sharma
Former Vice Chancellor
Sher-e-Kashmir University of Agriculture
& Technology of Jammu
Jammu (J&K), India

Sandeep Kumar
Former Employee
South Dakota State University
Brookings, SD, USA

ISBN 978-3-031-28056-6 ISBN 978-3-031-28057-3 (eBook)
https://doi.org/10.1007/978-3-031-28057-3

This Springer imprint is published by the registered company Springer Nature Switzerland AG
The registered company address is: Gewerbestrasse 11, 6330 Cham, Switzerland

Contents

Chapter 1
Soil Physical Productivity and Plant Growth

Plant growth and yield are the resultant of complex interactions among plant genotype, soil environment, climatic conditions and management practices. Every growing plant has unique and specific physiological requirements. Soil environment is complex and dynamic and essentially has three components, viz. physical, chemical and biological. All the three components are strongly interdependent and exhibit complex interactions. They influence plant growth in a variety of ways. The soil environment may be favourable or hostile to a given plant species. But because of large genetic variability among different plant species and the tremendous *plasticity*[1] of plants, one or the other plant species can grow successfully under each set of environmental conditions even though the full growth and yield potential of plant is not expressed. There is always an optimum environment under which plants can grow and perform optimally. This chapter defines and describes plant growth and plant growth stages, and their response to different environmental factors. The soil-physical heterogeneity at the macro and micro scale and its influence on plant root system are also discussed. It also describes soil physical productivity and its qualitative and quantitative evaluation. A brief reference is made to the soil physical resilience and plant productivity.

1.1 Plant Growth

The term *growth* refers to 'the progressive development of an organism'. The growth may also be defined as 'the irreversible permanent increase in size, volume or mass of the organism accompanied by an increase in its dry weight'. This definition of growth applies to all the living organisms.

[1] *Ability of plants to adapt themselves to varying environmental conditions.*

P. K. Sharma, S. Kumar, *Soil Physical Environment and Plant Growth*,
https://doi.org/10.1007/978-3-031-28057-3_1

The *plant growth* may be defined as 'the irreversible permanent increase in size, volume or mass of the plant accompanied by an increase in its dry weight'. The change in plant growth per unit time is known as plant *growth rate*. It is quantitative in nature.

Plant growth occurs in meristematic tissues, also called meristems. The meristems consist of actively dividing cells which are capable of forming new tissues. The meristems are found primarily at the growing tips of roots and shoots and in the cambium. Depending on their location in the plants, the meristems are classified as *apical* (located at root and shoot tips), *lateral* (located in the vascular and cork cambia), and *intercalary* (at internodes, and leaf bases). Apical meristems are responsible for growth of roots and shoots, while lateral meristems are responsible for increase in thickness and girth of stem. Different zones in a root tip of a vascular plant (plant having well developed root, shoot and vascular, i.e. xylem and phloem system) are shown in Fig. 1.1.

Many structural changes occur during plant growth, which are called the *developmental stages*. These developmental stages may be gradual (e.g. formation and maturation of tissues, formation of vegetative and floral buds) or abrupt (e.g. germination, flowering and senescence) and are qualitative in nature.

The complete life cycle of a plant, i.e. from seed to seed, may be divided into five growth phases, viz. germination, seedling emergence, vegetative phase, reproductive phase and maturity phase. Some workers also consider phases like seed storage, seed dormancy, seed dispersal, etc. as phases of plant growth cycle. Each growth phase responds differently to different environmental factors. The important environmental factors which affect plant growth are light, water, temperature, aeration and nutrition. If any of these factors is limiting, plant growth and performance are affected adversely.

The plant growth as a function of time, especially in case of annual plant species, follows a sigmoid curve (Fig. 1.2).

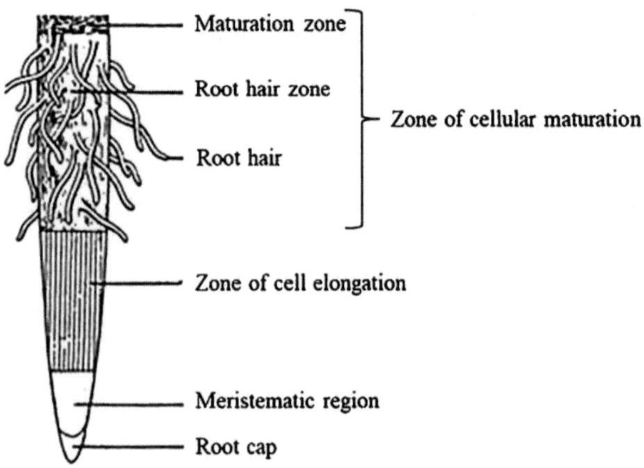

Fig. 1.1 Different zones in a root tip of a vascular plant

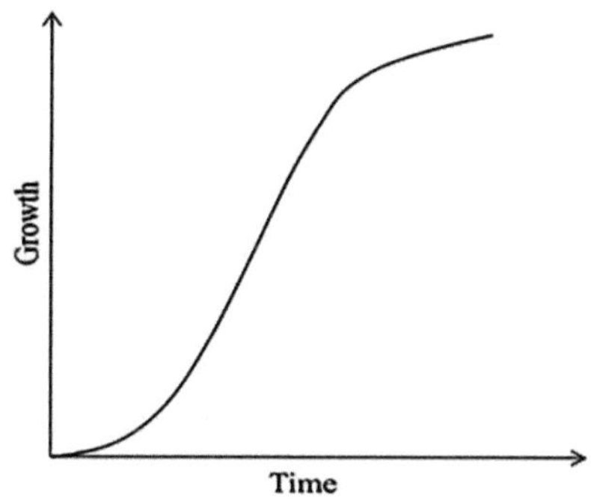

Fig. 1.2 The growth of an annual plant species as a function of time

1.2 Plant Growth Stages

The first step of plant growth is the seed germination, which depends on the seed viability. Hence, some workers consider seed dormancy also as a part of plant growth cycle. We shall also start our discussion with seed dormancy.

1.2.1 Seed Dormancy

All seeds remain viable for a specific period of time, and this viability period may range from a few weeks to many years depending on seed species and storage conditions. Lotus seeds have the maximum viability period of 1000 years. Unviable seeds fail to germinate.

In some crops, the freshly harvested seeds fail to germinate and require some time or need some treatment before they become capable of germination. Such seeds are called dormant seeds, and the process is known as *dormancy*. Almost all dehydrated seeds having moisture content around 6–15% are dormant. The seed dormancy may also be due to immature embryo at the time of harvest, lack of sufficient growth hormones, presence of growth inhibitors (e.g. abscisic acid), or presence of hard and impermeable seed coat. Dormancy of seeds may be broken in different ways:

- Storing seeds for some time for maturing of the embryo or synthesis of growth hormones
- Treating seeds with certain hormones like gibberellins
- Abrasion of the seed coat as is done in certain seeds of trees and shrubs
- Some seeds may require extended cold period, called *stratification period*
- Soaking of seeds in water before sowing

1.2.2 Germination

The *germination* may be defined as 'the process of emerging of a plant from its seed'. When seed is placed in an environment optimum for germination, there is activation of metabolic processes within the seed, resulting in the emergence of radical and plumule, and formation of a seedling. The process of germination starts with the absorption of water by the seed, followed by a series of biochemical reactions responsible for seed germination. During the germination period, the food reserve stored in seed is utilized. Hence, the process of germination ends with the appearance of radical when the seedling becomes capable of absorbing nutrition from soil.

1.2.3 Seedling Emergence

The word *emergence* is a noun and finds its origin in the Latin word *emergere,* meaning 'bring to light'. Seedling emergence, therefore, refers to 'the process of coming out of seedling from soil'. It follows the process of germination of seed.

1.2.4 Vegetative Phase

The *vegetative phase* refers to the growth period of a plant between germination and flowering. The plants during this phase synthesize and accumulate food through photosynthesis which is utilized during the flowering and reproductive growth phases.

1.2.5 Reproductive Phase

The *reproductive phase* refers to the growth stage at which plants produce flowers, fruits and seeds. The stage at which plants start flowering, the growth phase shifts from vegetative to reproductive phase.

1.2.6 Senescence Phase

The *senescence phase* refers to the growth stage at which plants show ageing and death. Senescence sets in after the seeds have matured, leading ultimately to the drying and death of the plant.

1.3 Environmental Factors Affecting Plant Growth

The environmental factors that influence plant growth can be classified into two main groups: (1) biotic and (2) abiotic factors.

1.3.1 Biotic Factors

Biotic factors are the 'living' components of environmental factors. They include a long list of macro- and microorganisms (flora and fauna, both). Macro-organisms include animals, humans, birds, insects, arachnids, molluscs and plants. Microorganisms include fungi, bacteria, virus and nematodes. The biotic factors may influence plant growth positively as well as negatively depending upon how they interact with the plants. These interactions include *mutualism*, *herbivory*, *parasitism* and *allelopathy*.

- *Mutualism*: It is symbiotic interaction between different species that is mutually beneficial (e.g. pollination).
- *Herbivory*: It refers to an act of eating plants. The animals that eat plants are called *herbivores,* e.g. cattle and goats. It has negative impact on plant growth.
- *Parasitism*: It refers to the relationship between two organisms in which one lives as a parasite on the other, e.g. insects and pests cause diseases in plants; weeds compete with plants for space, nutrients, water and light; root knot nematodes reduce absorption of nutrients from soil by plant roots. It has negative impact on plant growth.
- *Allelopathy*: It refers to the biochemical inhibition of one organism by the other. Harmful substances released by roots of some plants are toxic to other plant species due to allelopathy. It has a negative impact on plant growth.

1.3.2 Abiotic Factors

The abiotic factors are non-living components of environment and include topography, soil and climatic factors, viz. water, light, temperature, air, relative humidity and wind.

1.3.2.1 Topography

Topography refers to the physical features of the landscape, such as elevation, slope, terrain characteristics (flat, rolling, hilly, etc.), land forms (mountains, hills, valleys, lakes, rivers, dams, roads, etc.), and geographical location (latitude and longitude).

Elevation influences plant growth primarily through temperature effects. Generally temperature decreases by 1 °C for every 100 m rise in altitude in dry air (in moist air temperature drop is relatively low). Slope characteristics decide incidence of solar radiation, wind velocity, soil type and susceptibility of soil to erosion. All these parameters impact plant growth.

Topographic features are important in deciding crop species for a particular location, as the plant species are highly sensitive to abiotic factors.

1.3.2.2 Soil

Soil is the most important and vital component of the environment. It is a complex physical, chemical and biological system in nature. It is a store house of water and plant nutrients. It also stores plant pathogens and other phytotoxins. It supports and nourishes the plants by way of supplying to them physical support and essential growth factors like water, nutrients and oxygen. Pathogens and phytotoxins present in soil may harm the plants.

Soil serves as an excellent medium for plant growth. Most of the plant species are *terrestrial*, i.e. they require soil to grow, but there are plant species that do not require soil for their growth. They can grow in water (called *hydrophytes,* e.g. lotus, water lily, sea weeds, etc.) or on surface of another plants (called *epiphytes*, e.g. numerous ferns, bromeliads, air plants and orchids growing on tree trunks in tropical rainforests). Epiphytes are not parasites.

The effects of soil's physical, chemical and biological properties on plant growth are briefly described below. It may be noted that all soil properties and processes occurring in bulk soil, i.e. at macro-environment level are different from those in soil which is in close proximity to the root system, i.e. at micro-environment level, and plant growth responds differently to soil properties in these two environments.

(a) *Soil physical properties and plant growth:* Soil water, air, temperature and mechanical impedance (i.e. soil strength) are the four basic soil physical parameters that affect plant growth directly. They are known as *primary physical properties* of soil. All other commonly measured physical properties (and processes), viz. soil texture, soil structure, aggregation, aggregate stability, bulk density, porosity, pore-size distribution, infiltration, hydraulic conductivity, soil colour, etc., influence plant growth indirectly, i.e. through their influence on primary physical properties. They may be termed as *secondary physical properties*.

    Plants cannot grow without water. Soil must retain water in quantity and potential range that is available to plants. Too much or too little water stresses plant growth. Water as such is not toxic to plants. Excess water affects plant growth by restricting oxygen supply to roots. Adequate soil aeration is required for supplying O_2 to the respiring roots and plant metabolism, and for removal of excess CO_2 from soil atmosphere, which is a product of respiration and organic matter decomposition processes. All metabolic processes in plants and soil are temperature dependent, and hence soil temperature is an important factor in crop production. Soil mechanical impedance directly influences plant growth by affecting seedling emergence, root penetration and proliferation, and soil hydraulic properties and processes. Soil texture and soil structure play an important role in deciding soil's physical environment in relation to plant growth.

(b) *Soil chemical properties and plant growth:* Chemical properties of the soil (nutrient availability, cation exchange capacity, pH, salinity, etc.) are important components of soil fertility. The *soil fertility* refers to the plant nutrient status of soil. There are at least 18 elements which are essential for normal plant growth. The nine of them, viz. C, H, O, N, P, K, Ca, Mg and S are required in relatively large amounts by plants and are called *macro-nutrients.* The remaining nine elements, viz. Fe, Zn, Mo, Ni, Mn, B, Cu, Co and Cl, are needed by plants in much smaller quantities and are called *micro-nutrients* or *trace elements.* Except for C, H and O, which are largely found in air and water, all other nutrients occur in soil. Normally plants absorb all nutrients, except C, H and O, from the soil solution. Some elements, however, can also be absorbed by the leaves if sprayed on the foliage. For absorption from soil, the elements must occur in dissolved form in soil solution. Water and O_2 are two important environmental factors which help in dissolution and active absorption of nutrients. All nutrients must be available in balanced form for optimum plant growth and yield. The properties like pH and cation exchange capacity, etc. strongly influence nutrient solubility and retention in soil. Saline soils are unfit for normal growth of plants.

(c) *Biological properties and plant growth*: Biological properties include the living organisms (flora and fauna) and the organic matter in the soil. Living organisms found in a healthy soil vary from large creatures, such as earthworms, to the smallest bacteria, fungi and yeasts. Earthworms alone may weigh around 100 to >1500 kg/ha; fungi may weigh between 1000 and 15,000 kg/ha in a healthy soil. The population of these organisms in soil is favoured by high organic matter levels, adequate soil moisture, optimum temperature and good drainage and aeration.

The soil biology exhibits complex interactions with other soil properties (physical and chemical). Soil organisms help to decompose organic matter. The burrowing organisms incorporate the organic matter into the soil. They also create large pore spaces that improve soil aeration and infiltration. The smaller organisms, such as bacteria, actinomycetes, fungi, yeasts, algae and protozoa, further decompose the organic matter, which releases nutrients in a form that plants can use. They also have a significant impact on soil structure.

1.3.2.3 Climatic Factors

There are various climatic factors, viz. water, light, temperature, oxygen, relative humidity, winds, etc. that influence plant growth. These climatic factors always operate together and interact with each other under natural conditions.

1. *Water:* Water is the major constituent of plant tissues and most essential factor of plant growth. As high as 80–95% of fresh weight of most of the crop plants is water; fresh vegetables and fruits have relatively higher water contents. Water

supports cell structure, metabolic functions, photosynthesis, and transportation of nutrients and photosynthates in plants. Water is also an important factor for germination of seeds. The seeds must contain around 75–95% water to support metabolic processes within the seed embryo for seed germination. All dehydrated seeds are dormant, and they fail to germinate. Water is necessary for the activities of protoplasm, conversion of insoluble food into soluble form for its translocation to the embryo, transportation of dissolved oxygen for use by the growing embryo, softening of seed coat and improving permeability of seeds.

Precipitation (rainfall, snow, dew and hails) are the sources of water on earth. Rainfall is the single most important environmental factor that decides the distribution and performance of vegetation on the landscape. The rainfall characteristics (frequency, intensity, amount and duration) vary greatly with location and climate types. It has significant effect on the dominance of certain types of vegetation as well as crop growth and yield at different locations.

Depending on the water requirement, the plants may be grouped into three categories:

(a) *Hydrophytes:* Plants adapted to living in water or in water-saturated soils. The hydrophytes usually have large interconnected intercellular gas-filled spaces in their root and shoot tissues (called aerenchyma) to facilitate air exchange.
(b) *Mesophytes:* The most common terrestrial plants that are adapted to moist (i.e. neither too wet nor too dry) environments. The water requirement of these plants varies with the type of the plant (root system and other plant features).
(c) *Xerophytes:* Plants adapted to drought conditions. These plants have special features, such as reduced permeability to decrease water loss, swollen tissues to conserve water, or deep and extensive root systems to acquire water, to withstand long dry spells. The water requirement of these plants is relatively low.

Too much or too less water situation, as in water-logged and drought environments, respectively, is unfavourable for the normal growth of most of the (arable) crop plants.

2. *Light:* The light is essential for seed germination, chlorophyll production, photosynthesis, stomatal movement, phototropism, photomorphogenesis, translocation, mineral absorption and abscission, etc. Solar light has three characteristics which influence plant growth:

(a) *Intensity*: It refers to the quantity of solar radiation per unit time reaching the plant surface. It affects photosynthesis. Higher the light intensity better is the photosynthetic process, and vice versa. Light intensity is maximum during summers and minimum during winters.
(b) *Quality*: It refers to the colour or wavelength of solar radiation reaching the plant surface. Different wavelengths of solar radiation have different effects on plant growth. Green light is least effective, while red light is most effective in plant growth. Blue light primarily affects vegetative phase, i.e. leaf growth. A combination of red and blue lights encourages flowering in plants.

(c) *Duration*: Light duration, i.e. *photoperiod*, refers to the length of time for which a plant is exposed to sunlight. It affects primarily the flowering process of a plant. The length of uninterrupted dark period is critical to floral development. Based on their response to photoperiod, plants may be grouped into three categories, viz. short-day, long-day, or day-neutral plants. The *short-day plants* flower only when the day length is <12 h. The *long-day plants* flower only when day lengths exceed 12 h. The *day-neutral plants* flower regardless of day length.

3. *Temperature:* The temperature (of air and soil) influences all plant growth processes such as breaking of seed dormancy, seed germination, photosynthesis, respiration, transpiration, protein synthesis, translocation, etc. Relatively high temperatures favour the translocation of photosynthates in plants and cause the plants to mature earlier. Most of the plant species survive within a temperature range of 0–50 °C.

   All enzyme activities and biochemical reactions occurring within plants are temperature sensitive. Studies indicate that with every 10 °C rise in temperature within optimum temperature range, the enzymatic reactions are nearly doubled. At higher temperatures, denaturation of enzymes and proteins occurs and biochemical reactions are adversely affected. Similarly, excessively low temperatures are also harmful for plant growth and yield. For example, water absorption from soil is inhibited at low temperatures because water becomes more viscous, less mobile, and protoplasm becomes less permeable. At freezing temperatures, the cell walls may get ruptured, thereby adversely affecting plant growth.
4. *Oxygen:* Plants require energy for growth and this energy is released during the process of respiration. Plants require oxygen for respiration, which in case of all upland crops, is obtained from soil air. The germinating seeds respire very actively and require oxygen in sufficient quantities. In the absence of oxygen or due to its insufficient supply, the germination is adversely affected. That is why many seeds fail to germinate under waterlogged conditions or if placed deeper in soils where oxygen concentration is relatively low.
5. *Relative humidity:* The relative humidity (RH) refers to 'the amount of water vapours present in air relative to the maximum amount of water vapours than can exist in air at the given temperature'. For example, a 30% RH at 25 °C indicates that every kilogram of air contains 30% of the maximum water vapours that the air can hold at 25 °C. The RH is an indication of the relative wetness (or dryness) of air. Moist air has higher RH than dry air.

   The RH has significant effect on plant growth. It impacts photosynthesis and transpiration by regulating opening and closing of stomata. Dry air causes closure of stomata, thereby lowering the photosynthesis as well as transpiration. The RH also influences canopy temperatures by regulating transpiration process. Proper RH is maintained inside the propagation chambers for the proper propagation of plants from stem and leaf cuttings.
6. *Wind:* The wind, i.e. the air movement is the resultant of existence of pressure gradients caused by thermal variations in the air. Winds impact plant growth through their influence on evapotranspiration, photosynthesis, gas exchange

between atmosphere and soil, lodging of crops, etc. Winds also play an important role in pollination by way of acting as a vector of pollen grains from one location to other locations and from one flower to another flower.

1.4 Soil Physical Micro-environment: The Rhizosphere

The micro-environment refers to the narrow region of soil influenced by root activities. Lorenz Hiltner, a German agronomist and plant physiologist in 1904 (Hiltner 1904; Hartmann et al. 2008) observed that the properties of soil immediately in contact with the root system, i.e. at soil–root interface are different from those in bulk soil, and that plant growth is significantly affected by these properties. He termed this volume of soil as *rhizosphere*. This word is derived from the Greek word 'rhiza' which means 'root'. Since then the rhizosphere has been extensively studied and continues to be a hot research topic (Bakker et al. 2013). The above-ground plant foliage is called *phyllosphere* and is also a focus of interest for many plant researchers (Vorholt 2012).

The *rhizosphere* may be defined as 'the zone at the plant root-soil interface which is influenced by the root activities'. Rhizosphere is a micro-ecological zone in direct proximity of plant roots. Practically it refers to the soil that clings to roots after being gently shaken in water. The soil surrounding the rhizosphere is called *bulk soil*. The rhizosphere may be differentiated into three zones (McNear Jr 2013):

(a) The *endorhizosphere*: It includes portions of the cortex and endodermis in which microbes and cations can occupy the 'free space' between cells (called apoplastic space).
(b) The *rhizoplane:* It is the zone at the root–soil interface directly including the root epidermis and mucilage.
(c) The *ectorhizosphere*: It is the outermost zone of rhizosphere extending from the rhizoplane out into the bulk soil.

It is difficult to define precisely the shape and size of the rhizosphere because of the inherent diversity and complexity of the root systems. The actual volume of the rhizosphere depends on the root system architecture, zone of influence of the plant roots and associated microorganisms. The zone of influence of the plant roots in soil varies with the type of processes involved. For example, effects of root activities may be limited within 0.5 mm for nutrients with slow diffusion and up to more than 50 mm for gases. However, the most common effects are observed between 1 and 10 mm.

The physical, chemical and biological properties in the rhizosphere are different from that in the bulk soil. These properties show a radial as well as longitudinal gradient along the plant roots. The rhizosphere is a dynamic region in which several bio-geochemical processes (biological, geological and chemical) take place due to the physical activity of roots as well as root exudates and soil microorganisms. The rhizosphere, in general, has abundance and diversity of microbial population and is metabolically more active than the bulk soil. The microbial composition, their interactions, activities and processes in the rhizosphere (e.g. nutrient absorption, protection against pathogen attack, etc.) are important for root growth and plant health (Berendsen et al. 2012).

The properties of rhizosphere are influenced by the chemicals secreted by the roots and the associated soil microorganisms known as the *root microbiome*. The microbial abundance and diversity in the root microbiome are influenced by the factors like root system, soil properties and environmental conditions. Some of the factors are described below:

- *Root activities*: Plant roots release large amounts of carbon (10–250 mg C/g root produced or about 10–40% of their total photosynthetically fixed carbon) in the rhizosphere in organic (low and high molecular weight chemicals) and inorganic (e.g. HCO_3) forms (Newman 1985). The low molecular weight carbon compounds (e.g. organic acids, amino acids, proteins, sugar, phenolics and other secondary metabolites) are easily used by microorganisms. The high molecular weight carbon compounds (e.g. mucilage and cellulose), which make up the bulk of root-released carbon, are complex and insoluble molecules and are not easily used by the microorganisms. Death and decay of root tissues (root cap and border cells, etc.) also add substantial carbon to the rhizosphere. The composition and pattern of carbon deposits, called the *rhizodeposits*, significantly influence the chemical, physical and biological processes in the rhizosphere (Jones et al. 2009). The root exudates perform several functions in the rhizosphere, viz. acquisition of nutrients (e.g. Fe and P) by changing soil pH or redox potential or chelating the nutrient elements in the rhizosphere, invasiveness (i.e. allelopathy) or as chemical signals to attract symbiotic partners (chemotaxis) (e.g. rhizobia and legumes) or the promotion of beneficial microbial colonization on root surfaces (e.g. *Bacillus subtilis*, *Pseudomonas florescence*) (Bais et al. 2004). Dennis et al. (2010) suggested that the influence of root exudates on rhizosphere bacterial communities is more than the influence of other rhizosphere carbon pools.
- *Plant species*: The root exudates are plant-species specific (Rovira 1969). Hence, microbial diversity differs in the rhizosphere of different plant species. Plant genotypes are selective in attracting bacterial population from the bulk soil. Also, when plants are under attack of pathogens, they selectively attract bacteria from surrounding bulk soil that protect plants against the pathogen attack.
- *Soil properties*: The properties of bulk soil directly influence the microbial population as well as the rates of processes they perform in the rhizosphere. For example, microbial activity increases with temperature, which in turn affects rates of decomposition of organic matter. Microbial processes, on the other hand, directly affect their environments, contributing to the carbon and nitrogen cycles, which are important for microbial and plant health. At the micro scale, bacteria and other microbes participate in a variety of reactions that affect nutrient cycling, pH, as well as oxygen and CO_2 content.
- *Drought stress*: Certain bacteria are capable of increasing photosynthesis and biomass production in plants under drought stress conditions (Marasco et al.

2012). Hence, rhizosphere microbiome under drought stress environments may have higher population of such bacteria.

Plants generally shape their rhizosphere microbiome to their own advantage (Bakker et al. 2013). Plant roots secrete water soluble compounds (e.g. sugars and amino acids) and microorganisms feed on them. It leads to the symbiotic relationships between plants and microorganisms. Sometimes plant root-microorganism association in the rhizosphere may be harmful also to the plant growth and health. Plant roots selectively attract microorganisms from the bulk soil into the microbiome and also influence microbiomes's specific functions like antibiotic production, biofilm formation, conjugation, motility, symbiosis, and virulence (Miller and Bassler 2001). The association between plant roots and microbiome is very complex. An understanding of the processes and mechanisms operating in the rhizosphere may help in managing soils to improve plant productivity and environmental quality.

1.5 Soil Physical Heterogeneity and Root Growth

Soil system is highly complex and heterogeneous with respect to physical, chemical and biological properties at spatial and temporal scales. At larger spatial scales, soil heterogeneity may influence the distribution and productivity of plants in agro-ecosystems, while at smaller spatial scales it may affect root systems, and morphological and physiological plant responses. Over the last about two decades, several studies have been published on plant responses to soil heterogeneity (Garcia-Palacio et al. 2012).

Soil physical properties under field conditions are highly variable in space and time. Their precise characterization depends on the scale of sampling. Such variations, coupled with chemical and biological properties, have a significant impact on the growth of plants. Root growth and morphology (root length, thickness and branching) show significant response to soil heterogeneity.

The root distribution pattern within the soil profile through space and time is known as *root architecture* (Lynch 1995). The plant rooting pattern, called the *root system architecture* (RSA), is significantly affected by soil properties, plant species and climatic conditions. Root architecture is governed by three root processes, known as 'component traits', viz. elongation rates, branching rates and tropism, i.e. root growth response (positive or negative) to the external stimuli like gravity (gravitropism or geotropism), light (phototropism), water (hydrotropism), nutrients and mineral toxicity (chemotropism), etc. The root system architecture is influenced primarily by three factors, viz. plant's genetics, gravitational force, soil conditions (mechanical impedance, water, nutrients, aeration, soil temperature etc.) and environment. Gravity and genetic factors are static while soil and environmental conditions are variable over time and space. Rich and Watt (2013) have extensively reviewed the effects of soil conditions and plant genetics on root growth patterns of cereal crops.

Higher mechanical impedance restricts root growth. It decreases elongation of primary roots but increases root thickness; its effect on growth of lateral roots is not clearly understood (Rich and Watt 2013). Some studies did not find any effect of soil strength on the density of lateral roots (number per unit main axis length) in barley, wheat and oats, while others observed significant effect of soil strength on branching density. Some studies observed acceleration in the initiation of branch roots in dense soils compared with those in loose soils. Studies have also shown that hardpans can decrease rooting depth but not necessarily the root mass because plants may respond with increased root exploration above the hardpan. In compact soils, enhanced root growth is observed in root channels and soil cracks. The tendency of roots to explore soil cracks, root channels and macro pores may allow the roots to penetrate hardpans or to explore deeper depths in soil profiles. Strong genotypic differences in root penetration ability have been observed in cereal crops.

1.6 Soil Physical Productivity

Soil productivity is defined as 'the capacity of soil to produce crops under a specified management system'. Soil productivity is governed by soil properties (physical, chemical and biological) and management practices (viz. tillage, irrigation & drainage, nutrient management, crop rotations, pest & disease management, etc). A productive soil is one in which physical, chemical and biological conditions are favourable for crop production. Based on the nature of soil property, soil productivity may be resolved into *physical productivity*, *chemical productivity* and *biological productivity*.

Soil physical productivity is a reflection of soil physical quality. *Soil physical quality* is 'the ability of a soil to meet the plant and ecosystem requirements for water, aeration and strength over time, and to resist and recover from processes that might diminish that property'.

1.7 Evaluation of Soil Physical Productivity

Compared to chemical and biological productivity, physical productivity is less understood. It has not been possible to precisely ascertain critical values of physical properties and processes optimum for plant growth. There are several reasons for it. Some of them are enlisted below:

- *Soil physical environment is highly complex and heterogeneous at spatial and temporal scales.* The plant growth responds to soil heterogeneity differently at different spatial and temporal scales. For example, soil heterogeneity at smaller spatial scales may affect morphological and physiological plant responses, while at larger spatial scales may influence the distribution and productivity of plants in agro-ecosystems. Over the last about two decades, several studies have been published on plant responses to soil heterogeneity (Garcia-Palacio et al. 2012).

- *A complex interaction exists among environment, soil properties—physical, chemical and biological, and plant growth.* It is difficult to develop simple correlations among soil physical properties and plant growth having universal/large scale applications, although important generalizations can be made. The information generated is usually site specific and restricted to a given agro-ecological situation.
- *Soil physical properties are interdependent.* It is difficult to change one property without changing the other properties. For example, any change in soil water content changes soil water potential, soil aeration, soil temperature and soil mechanical impedance. The effect of one soil property on plant growth varies with the change in the other soil property(ies). For example, the adverse effect of higher mechanical impedance on growth of chickpea (*Cicer arietinum*) is reduced at higher oxygen diffusion rates (Rickman et al. 1966). Consequently, it is nearly impossible to isolate the effect of a single factor on the plant growth.
- *The methodology of assessing long-term plant responses to minor physical differences is lacking.* It makes the assessment of plant responses to adversities during a plant's life cycle difficult. It is generally difficult in edaphological experiments to discriminate between small differences, say of the order of <5–10%, which otherwise may have significant effect on plant growth when considered over the plant's complete life cycle.

Nevertheless, several workable criteria, *qualitative* and *quantitative*, for the assessment of soil physical productivity have been developed. The choice depends on the purpose of study and availability of resources. The qualitative evaluation involves field observations, while quantitative evaluation integrates different soil physical parameters determined in the laboratory or field conditions.

1.7.1 Qualitative Evaluation of Soil Physical Productivity

Field observations can give some idea about the physical productivity status of soil:

- *Lack of vegetation or patchy growth.* It is an indicator of degraded soil structure or poor infiltration resulting from non-wetting soil characteristics or subsoil impermeability.
- *Weedy vegetation.* Certain weeds, such as *Cyperaceae* or *Juncaceae*, are indicators of water logging, which may result from poor infiltration or existence of an impermeable subsurface horizon or shallow water table.
- *Rill and sheet erosion* or *turbidity of water in ponds and lakes after rain* indicate poor surface soil structure.
- *Surface crusts and hard-setting surfaces* are indicators of unstable surface soil structure.
- *Poor infiltration and ponding* are indications of poor soil structure.

- *Pale surface soil colour* and *absence of organic matter* indicate degraded soils, low organic matter content and loss of clay either through eluviation[2] or by water or wind erosion.
- *Cloddiness* upon tillage indicates compact and dry seed-bed.
- *Restricted root growth* is an indication of soil compaction or presence of compact/hard subsurface soil layers.

1.7.2 Quantitative Evaluation of Soil Physical Productivity

Determination of soil physical properties and correlating them with plant growth, yield and yield components is a simple, easy and commonly used approach of assessing soil physical productivity. Several such correlations have been developed and reported in literature. The conclusions drawn are site specific and are applicable under a given agro-ecological situation. Such studies fail to develop meaningful critical values of soil physical properties in relation to plant growth which have universal application because of heterogeneity, strong inter-dependence and complex interaction among different soil physical properties (Sect. 1.3). Large spatial and temporal variability in soil physical properties makes development of large data sets to develop meaningful correlations difficult.

The problems of developing large data sets may be overcome to some extent by using *pedotransfer functions*. The pedotransfer functions are predictive functions for some soil attributes using available minimum data set without actually measuring the specific soil attributes. The term pedotransfer function was coined by Johan Bouma (Bouma 1989) for translating 'data we have' into 'what we need'. These functions are easy to develop. For example, changes in soil organic matter content can be estimated by using simple regression equations based on the amount of crop residues added to soil. Likewise, soil hydraulic properties including water retention, hydraulic conductivity, water holding capacity, etc. can be estimated from the data on soil texture and organic matter content. With the help of these functions, changes in many important soil attributes can be simulated from relatively few measured indicators of soil quality.

Several attempts have been made to develop soil indicators that are simple and have a physical meaning in relation to soil physical environment and plant growth. Any index evaluating soil physical productivity must incorporate all or most of the critical measurable sources of stress that the soil imposes on growing plants. Many 'single value soil physical indices' based on few soil physical parameters have been developed by different workers, viz. '*least limiting water range*' (Letey 1985; da Silva et al. 1994), '*tilth index*' (Singh et al. 1992), 'soil *physical health index*' (Gupta and Abrol 1993), '*soil physical quality index*' (Dexter 2004), '*soil quality index*' (Mohanty et al. 2006), etc. Each index has its own advantages and limitations. Some of the indices have been briefly described below.

---

[2] Removal of dissolved or suspended material from a soil layer by the downward movement of water.

1.7.2.1 Least-Limiting Water Range (LLWR)

Letey (1985) introduced the concept of '*non-limiting water range*' (*NLWR*) as a single value soil physical index. da Silva et al. (1994) preferred to call this index as '*least-limiting water range*' (*LLWR*) rather than '*non-limiting water range'* (*NLWR*) with the argument that under field conditions it is practically impossible to get soil water range wherein soil physical conditions are 'non-limiting' to plant growth. A 'least limiting' water range is possible wherein soil physical constraints to plant growth are minimal. Moreover, plant growth varies in a continuous fashion with changes in soil physical properties rather than following a step function as indicated by *NLWR*.

The *LLWR* refers to 'the soil water range wherein limitations for plant growth associated with water content, aeration and mechanical impedance are minimal'. The index is based on three primary soil physical properties, viz. soil water, soil air and soil mechanical impedance.

The upper limit of *LLWR* is the water content at which soil air is not limiting. This value could be the water content at critical oxygen diffusion rate, i.e. 20×10^{-8} g/cm^2-min ODR (Lemon and Erickson 1952) or 10% air-filled porosity or −10 kPa matric potential. The lower limit is the water content below which soil mechanical impedance becomes critical for root growth (i.e. root growth decreases by half). For most of the cereal crops, the critical soil penetration resistance is around 1.75–2.00 MPa. Hence, the *LLWR* may be computed as the difference between water contents (by volume) at 10% air-filled porosity and at 1.75–2.0 MPa soil penetration resistance, respectively. Higher the value of *LLWR*, better is the soil physical condition for plant growth, and vice versa. The conceptual relationship among bulk density, *LLWR* and plant available water capacity (*PAWC*) has been shown in Fig. 1.3.

The *LLWR* index has the advantage of integrating three primary soil physical properties which directly affect plant growth, viz. soil water, soil air and mechanical impedance in the root zone. In two long-term field experiments, one each under rice-wheat and maize-wheat system in a silty clay loam soil, wheat grain yields were significantly correlated with *NLWR* (Figs. 1.4 and 1.5). da Silva and Kay (1967a, b) demonstrated that crops growing in soils having small values of *LLWR* are relatively more vulnerable to, both, drought and high precipitation situations.

To determine critical value of this index across the soil textures, Topp et al. (1994) suggested the use of *LLWR: PAWC* ratio instead of *LLWR*. The hypothesis is that in a physically healthy soil, the *LLWR* should be equal to plant available water capacity (PAWC) (see section 'A' in Fig. 1.3) and the *LLWR: PAWC* ratio becomes 1.0. In this soil, the growing plants experience minimum constraints, if any, with respect to aeration and soil mechanical impedance over the entire available water range (i.e. *PAWC*). A smaller ratio is an indication of poor soil physical health and production potential. In two long-term field experiments conducted by Sharma and Bhushan (2001) and Verma and Sharma (2008) in a silty clay loam soil in India, the beneficial effects of organic residue recycling on soil physical productivity were clearly indicated by *NLWR:PAWC* ratio (Tables 1.1 and 1.2).

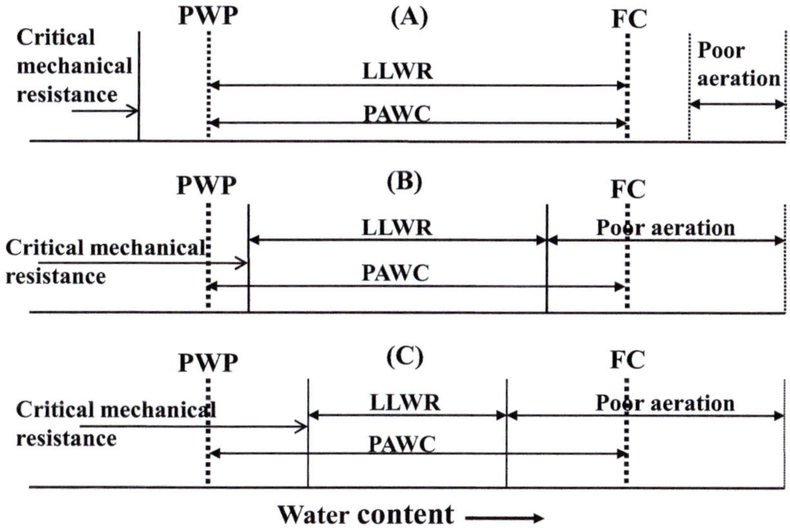

Fig. 1.3 A generalized relationship between least-limiting water range (LLWR), plant available water capacity (PAWC) and soil physical properties (*FC* field capacity, *PWP* permanent wilting point) (adapted from Letey 1985) (Note: Bulk density of soil increased from A to C)

Fig. 1.4 Relationship between NLWR and grain yield of wheat in a long-term field experiment under rice-wheat system (Sharma and Bhushan 2001)

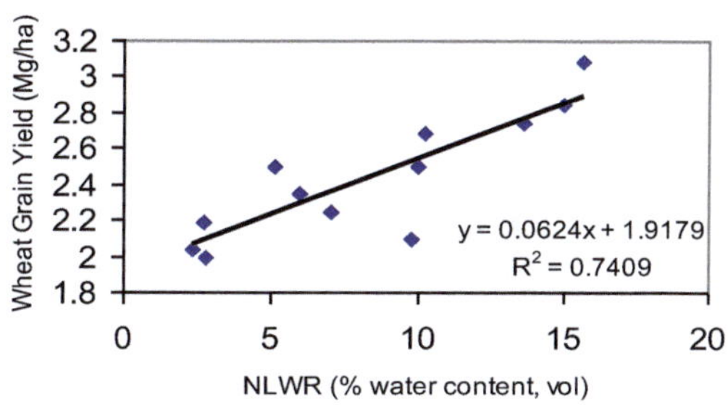

Fig. 1.5 Relationship between NLWR and wheat grain yield in a long-term field experiment under maize-wheat system (Verma and Sharma 2008)

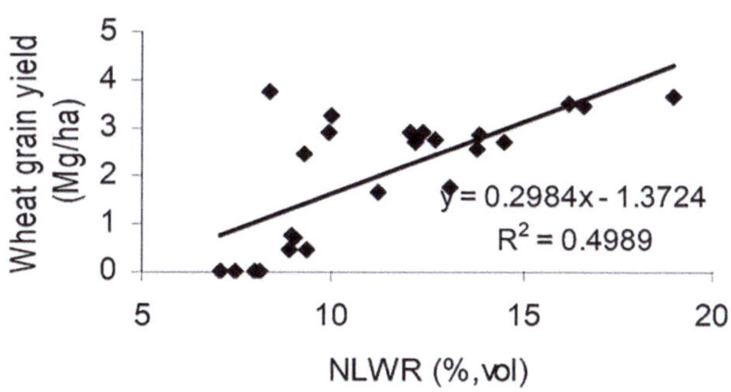

Table 1.1 Effect of ten annual additions of *lantana* biomass on plant available water capacity (*PAWC*), non-limiting water range (*NLWR*) and *NLWR:PAWC ratio* (adapted from Sharma and Bhushan 2001)

Lantana biomass (Mg/ha)	*PAWC* (%, vol)	*NLWR* (%, vol)	*NLWR:PAWC* ratio
0	12.9	4.3	0.33
10	13.4	7.4	0.55
20	13.7	10.9	0.80
30	14.9	15.1	1.01
LSD (0.05)	1.1	1.0	0.17

Table 1.2 Long-term effect of organic residues on plant available water capacity (*PAWC*), non-limiting water range (*NLWR*) and *NLWR:PAWC* ratio (adapted from Verma and Sharma 2008)

Cropping system	Duration (years)	Treatment	*AWC* (%, vol)	*NLWR* (%, vol)	*NLWR:PAWC* ratio
Maize-wheat	32	100%NPK	17.1	13.1	0.77
		100%NPK + FYM[a]	18.0	16.5	0.92
Rice-wheat	18	100%NK	14.1	7.5	0.54
		100%NK + lantana[b]	15.7	12.7	0.82

[a] 10 Mg/ha FYM
[b] 40 Mg/ha fresh lantana biomass

1.7.2.2 Tilth Index (TI)

Singh et al. (1992) developed a '*tilth index*' (*TI*) to quantitatively assess the quality of soil tilth which is an indicator of soil physical productivity. It is based on five soil physical properties, viz. bulk density, cone index, aggregate size uniformity coefficient, organic matter and plasticity index. The index value ranges between zero for conditions unusable by the plant and 1.0 for a soil that is non-limiting for plant growth. The *TI* was evaluated in field experiments near Ames, Iowa and Waseca and Minnesota (USA). The *TI* was positively correlated with yields of corn and soybean.

Emadi et al. (2008) observed *TI* as a sensitive index to changes in soil physical properties caused by changes in land-use pattern in North Highlands of Iran. Three adjacent land-use types including croplands (converted from pastures and forests for 18 years), a native forest and pasture lands were investigated. The *TI* decreased approximately 40% after conversion of native forest and pasture to cropland (Table 1.3).

1.7.2.3 Soil Physical Health Index (PI)

Gupta and Abrol (1993) developed '*soil physical health index*' (PI) for assessing the soil physical health status and its production potential. The index is based on eight parameters, viz. soil depth, water table depth, slope, bulk density, saturated hydraulic conductivity, available water retention capacity, organic carbon and non-capillary

Table 1.3 Comparison of tilth index (*TI*) of forest, pasture and cropland soils (adapted from Mostafa et al. 2008)

Land use	*TI* at soil depth (cm)		
	0–10	10–20	0–20
Forest	0.864[a]	0.868[a]	0.866[a]
Pasture	0.844[a]	0.842[a]	0.843[a]
Cropland	0.619[b]	0.524[b]	0.596[b]

Means followed by a same letter are not significantly different ($p < 0.05$)

Table 1.4 Physical health status and production potential of soils based on their 'soil physical index' (PI) (Gupta and Abrol 1993)

Class	Soil physical index (PI)	Physical health and production potential of soil
I	>0.75	Very good
II	0.50–0.75	Good, expected yield 75% of potential
III	0.25–0.50	Medium, expected yield 50% of potential
IV	<0.25	Poor, low yields, unsuitable for economical crop production

pores. Each of these parameters is assigned a score value. The maximum assigned score is 1.0 if the parameter value lies within the optimum range. If the value is below or above the optimum range, the score is less than 1.0. Higher the deviation of the parameter from the optimum range, lower is the score assigned to it.

The PI is computed as a product of these eight parameters, viz. soil depth (A), water table depth (B), slope (C), bulk density (D), saturated hydraulic conductivity (E), available water retention capacity (F) organic carbon (G) and non-capillary pores (H).

$$\mathrm{PI} = \mathrm{A}\cdot\mathrm{B}\cdot\mathrm{C}\cdot\mathrm{D}\cdot\mathrm{E}\cdot\mathrm{F}\cdot\mathrm{G}\cdot\mathrm{H} \tag{1.1}$$

Based on PI value, the soils are classified into four categories with respect to their physical health status and production potential, viz. very good, good, medium and poor (Table 1.4).

1.7.2.4 Soil Physical Quality Index (S)

Dexter (2004) proposed a soil *physical quality index* (*S*). The physical quality (*S*) index is equal to 'the slope of the soil water retention curve at its inflection point'. Larger values of S are associated with less compaction, higher soil organic matter (SOM) contents, higher micro-porosity and better microstructure. The *S* is a better index of soil as a rooting medium than bulk density. A value of S = 0.035 is the boundary between good and poor soil physical conditions. The index is found to decrease with increasing bulk density, but appears to be independent of soil texture.

1.7.2.5 Soil Quality Index (SQI)

Mohanty et al. (2006) developed a *soil quality index* (SQI) by regressing bulk density, penetration resistance, water stable aggregates and soil organic matter with crop yields (rice and wheat) for evaluating soil quality under rice-wheat system. According to these workers, the optimum SQI values for rice, wheat and rice + wheat yields were 0.84–0.92, 0.88–0.93 and 0.86–0.92, respectively. They also used SQI to understand the sustainability of rice-wheat system under different tillage and organic residue management practices.

Biswas et al. (2017) evaluated soil quality in rice-rice system on farmers' fields under three soil orders, viz. Inceptisols, Entisols and Alfisols in sub-tropical India using SQI. The SQI indicators were identified through principal component analysis using data of 37 physical, chemical and biological properties. Inceptisols showed better quality than Entisols and Alfisols.

1.8 Soil Physical Resilience and Plant Productivity

1.8.1 Soil Resilience

Whenever a soil is subject to any 'disturbance', either of the two things can happen. The soil either shows a resistance to any kind of change or if the change occurs, the soil tries to return to the original state or to a new dynamic equilibrium state. These two processes are collectively called 'soil resilience'. Some investigators, however, distinguish between these two processes by using two different terms, viz. 'soil resistance' for soil's capacity to resist change and 'soil resilience' for soil's ability to return to the original/new equilibrium state. *Soil physical resilience* refers to the ability of soil to resist changes in soil physical properties and processes.

Disturbance is defined as 'any event, natural or anthropogenic, that causes a significant change from the normal pattern or functioning of an ecosystem'. Natural disturbances include fires, earthquakes, floods, landslides, high intensity storms, etc. Anthropogenic disturbances include land use and land management, tillage, irrigation, drainage, grazing, cropping systems, farm mechanization, etc.

Several other terms are also in use with reference to soil resilience process. Different terms used in soil resilience process are summarized as follows (Lal 1997):

1. *Resilience*, the ability of a soil to resist adverse changes under a given set of ecological and land-use conditions and to return to its original or new dynamic equilibrium after disturbance
2. *Resistance*, the ability to resist displacement from the antecedent state or the capacity of a soil to continue to function without change throughout the disturbance
3. *Elasticity*, the rate of recovery after disturbance
4. *Amplitude*, the range of change in a property from which recovery is possible
5. *Hysteresis*, the divergence in the recovery path or pattern
6. *Malleability*, the difference in the new versus the antecedent state

1.8.1.1 Basic Components of Soil Resilience

Soil resilience has the following three basic components:

1. Tolerance against stress—environmental or anthropogenic
2. Ability of a soil to resist adverse changes under a given set of ecological and land-use conditions, and to return to its original dynamic equilibrium after disturbance
3. Ability of soil to return to a new dynamic equilibrium after disturbance

1.8.1.2 Soil Resilience Processes

Principal processes involved in soil resilience are (Biswas 2016) as follows:

1. Control of soil organic matter
2. Improvement in soil structure
3. Increase in biodiversity
4. Reduction in soil degradation and erosion rates below the soil formation rate
5. Increase in nutrient status and recycling mechanisms

1.8.2 Evaluation of Soil Resilience

Soil resilience is an important indicator of soil quality. Attempts have been made to characterize soil resilience based on soil quality. One such mathematical expression of soil resilience is as follows (Manna and Sharma 2009):

$$\text{Soil resilience} = f\,(\text{SQ, land use management})\ t \tag{1.2}$$

and

$$\text{SQ} = f\ \left(W_c \times S_c \times R_d \times E_d \times N_c \times B_d\right)\ t \tag{1.3}$$

where SQ is the soil quality, W_c the water capacity or the least-limiting water range, S_c the structural index, R_d the rooting depth, E_d the charge density or CEC, N_c the nutrient supplying capacity, B_d soil biodiversity, and t the time.

Biswas (2016) developed a *soil resilience index* (RI) based on the *soil quality index* (SQI).

$$\text{SQI} = \sum_{i=1}^{n} W_i S_i$$

where W_i is the weighing factor and S_i is the score of the prescribed variable. The value of SQI is used to compute soil resilience index (RI) as follows:

$$\mathrm{RI} = \left[\frac{\{\mathrm{SQI}(T) - \mathrm{SQI}(d)\}}{\{\mathrm{SQI}(p) - \mathrm{SQI}(d)\}} \right] \times 100 \tag{1.4}$$

where SQI(*T*) is the soil quality value after management intervention (*T*), SQI(*d*) is the soil quality value before management intervention (*T*), and SQI(*p*) the computed SQI value of the pristine (i.e. undisturbed, virgin) soil near the corresponding site. The numerator in Eq. (1.4) indicates recovery in soil quality due to management intervention and the denominator indicates the loss in soil quality due to degradation processes. The RI, thus, may be used for assessing soil management sustainability.

1.8.3 Factors Affecting Soil Resilience

- Soil resilience depends on *soil properties* like soil texture (especially the clay content), soil depth, SOM content and iron oxides. Deep soils, rich in clay, SOM and iron contents have better resilience. That is why Oxisols are more resilient than the Ultisols (de Andrade et al. 2017).
- *Climatic factors*, such as precipitation, temperature, radiation, relative humidity, evapotranspiration, etc., influence soil resilience. Soil resilience is influenced by the balance between soil degradation and restructuring, wet-dry cycles, freezing and thawing, biological activity and root growth (de Andrade Bonetti et al. 2017).
- *Soil management practices* that improve SOM status and maintain stable soil structure, such as conservation tillage, organic residue recycling, improving soil drainage, growing cover crops, sod culture, etc. increase soil resilience.

Soil resilience is a time dependent process. Soils may take variable times to recover to the original or the new equilibrium state. The mechanisms involved in soil resilience generally decide such recovery processes.

1.8.4 Significance of Soil Resilience in Agriculture

Soil resilience is an indicator of soil quality. Soil (physical) resilience is important for sustaining production systems. Since, resilient ecosystems are less vulnerable to land-use modifications, they offer an opportunity to adjust production systems according to the expected changes in climatic patterns and associated processes.

Lal (1997) classified soils into five categories depending on their recovery rates after disturbance (Table 1.5).

Table 1.5 Soil resilient classes (Lal 1997)

Class	Resilience	Description
0	Highly resilient	Rapid recovery, high buffering
1	Resilient	Recovery with improved management
2	Moderately resilient	Slow recovery with high input
3	Slightly resilient	Slow recovery even with change in land use
4	Non-resilient	No recovery even with change in land use

Highly resilient soils have high buffering capacities, high rates of recovery, and large amplitudes. They can perform their functions efficiently and profitably after recovery from disturbance or in the new state attained after disturbance. Non-resilient soils are fragile, unstable, unable to recover to the initial state, loose some or all of their specific functions in the new state, and become dysfunctional following disturbance. Soil resilience can be improved through management practices.

1.9 Question Bank

1.9.1 Short Questions:

(i) Define plant growth. Enlist important growth stages/phases of a plant growth cycle.
(ii) Differentiate between seed viability and seed dormancy.
(iii) In what way biotic factors affect plant growth?
(iv) Differentiate between primary and secondary soil physical properties in relation to plant growth.
(v) In what way light impacts plant growth?
(vi) What is rhizosphere? In what way plant growth influences the properties of rhizosphere?
(vii) What is meant by 'root system architecture (RSA)'? What factors influence RSA?
(viii) Is it possible to judge soil physical constraints simply from visual field observations? Justify your answer with examples.
(ix) What is 'least-limiting water range (LLWR)'? How to calculate this index?
(x) What is soil physical productivity?
(xi) Explain soil rhizosphere. How is it different from the bulk soil?
(xii) What factors influence the volume of rhizosphere?
(xiii) What is root microbiome? Briefly describe factors affecting the properties of root microbiome.
(xiv) What is the significance of root exudates in the rhizosphere?
(xv) Why is it difficult to quantitatively evaluate soil physical productivity?

(xvi) What is soil physical indexing? Briefly explain the principle of 'Least-Limiting Water Range'.
(xvii) What are pedotransfer functions? What is their significance in indexing soil physical health?
(xviii) What is soil resilience? What are three basic components of soil resilience?
(xix) Briefly explain 'soil resilience index'.
(xx) Briefly explain significance of soil resilience in agriculture?
(xxi) How will you characterize soil resilience? Briefly describe various factors affecting soil resilience.

1.9.2 Explain Why?

(i) Freshly harvested seeds of some crops fail to germinate and require some time or need some treatment before they become capable of germination.
(ii) Too much water stresses plant growth.
(iii) Soil temperature is an important abiotic factor in crop production.
(iv) Adequate light intensity is important for optimum plant growth.
(v) The seed germination process is highly temperature sensitive.
(vi) Many seeds fail to germinate at temperatures >45 °C.
(vii) Many seeds fail to germinate in waterlogged soils.
(viii) It is difficult to define precisely the shape and size of the rhizosphere.
(ix) Microbial diversity differs in the rhizosphere of different plant species.
(x) Rhizosphere microbiome under drought stress environments may have higher population of certain bacteria.
(xi) It is difficult to isolate and decipher the impact of a single soil factor on plant growth.
(xii) 'Least limiting water range' (LLWR) is a relatively better single value soil physical index
(xiii) Rhizosphere is metabolically more active than the bulk soil.
(xiv) Root activities have significant effect on the properties of rhizosphere.
(xv) Plants generally shape their rhizosphere microbiome to their own advantage.
(xvi) It is difficult to precisely ascertain critical values of physical properties and processes optimum for plant growth.
(xvii) Pedotransfer functions are useful tools in developing soil physical indices.
(xviii) Soil resilience is important trait of a productive soil.
(xix) Oxisols are more resilient than the Ultisols.
(xx) SOC significantly affect soil resilience.
(xxi) Soil resilience is an indicator of soil quality.

1.9.3 Fill in the Blanks:

(i) Plants possess tremendous .... by virtue of which they can adapt themselves to varying environmental conditions.
(ii) Plant growth occurs in .... tissues.
(iii) Freshly harvested seeds of some crops fail to germinate because of ....
(iv) Almost all dehydrated seeds having moisture content around .... are dormant.
(v) .... is a hormone used to treat seeds to break their dormancy.
(vi) The growth stage at which plants show ageing and death is called ....
(vii) *The biochemical inhibition of one organism by the other is called* ....
(viii) The symbiotic interaction between different species that is mutually beneficial is called ....
(ix) The plants which grow in water are called ....
(x) The plants which grow on the surface of other plants are called ....
(xi) The plants adapted to drought conditions are called ....
(xii) The most common terrestrial plants that are adapted to moist environments are called ....
(xiii) The seeds must contain around .... percent moisture to support metabolic processes within the seed for their germination.
(xiv) Green light is .... in plant growth.
(xv) Red light is .... in plant growth.
(xvi) The long-day plants flower only when day length exceeds ....
(xvii) Most of the seeds fail to germinate at temperatures ....
(xviii) The narrow region of soil influenced by root activities is known as ....
(xix) The plant rooting pattern in soil is called the ....
(xx) The ability of a soil to resist adverse changes under a given set of ecological and land-use conditions and to return to its original or new dynamic equilibrium after disturbance is known as ....
(xxi) The ability of a soil to meet the plant and ecosystem requirements for water, aeration and strength over time is called ....
(xxii) The .... is the zone at the plant root-soil interface which is influenced by the root activities.
(xxiii) The .... refers to the above-ground plant foliage.
(xxiv) The outermost zone of rhizosphere extending from the rhizoplane out into the bulk soil is called ....
(xxv) Plant roots release .... of carbon in the rhizosphere.
(xxvi) The zone of influence of plant roots in soil is generally observed between ....
(xxvii) .... is an example of 'primary soil physical property'.
(xxviii) Bulk density is a .... soil physical property.

(xxix) The …………… Is used for translating 'data we have' into 'data what we need'.
(xxx) The upper limit of LLWR is the water content at which ODR is …………
(xxxi) The lower limit of LLWR is the water content at which soil penetration resistance is …………
(xxxii) The value of 'Tilth Index' developed by Singh et al. (1992) varies between ………
(xxxiii) A value of 'soil physical health index' developed by Gupta and Abrol (1993) of …….. indicates very good soil physical health for plant growth.
(xxxiv) A value of 'soil physical quality index' developed by Dexter (2004) of …….. is the boundary between good and poor soil physical condition.
(xxxv) The ability of a soil to resist adverse changes under a given set of ecological and land-use conditions and to return to its original or new dynamic equilibrium after disturbance is called …………
(xxxvi) The rate of recovery in soil health after disturbance is called ……….
(xxxvii) The ability of soil to resist displacement from the antecedent state is known as ………

[**Key:** (i) Plasticity, (ii) Meristematic, (iii) Dormancy, (iv) 6–15%, (v) Gibberellins, (vi) Senescence phase, (vii) *Allelopathy, (viii)* Mutualism, (ix) Hydrophytes, (x) Epiphytes, (xi) Xerophytes, (xii) Mesophytes, (xiii) 75–95%, (xiv) Least effective, (xv) Most effective, (xvi) 12 h, (xvii) >45 °C, (xviii) Rhizosphere, (xix) Root system architecture, (xx) Soil resilience, (xxi) Soil physical quality, (xxii) Rhizosphere, (xxiii) Phyllosphere, (Xxiv) Ectorhizosphere, (xxv) 10–250 mg C/g root, (xxvi) 1–10 mm, (xxvii) Soil water or soil air or soil temperature or soil strength, (xxviii) Secondary, (xxix) Pedotransfer function, (xxx) 20 × 10^{-8} g/cm^2-min, (xxxi) 1.75–2.0 MPa, (xxxii) 0 and 1.0, (xxxiii) >0.75, (xxxiv) 0.035, (xxxv) Soil resilience, (xxxvi) Soil elasticity, (xxxvii) Soil resistance].

1.9.4 State Whether the Following Statements Are True (T) or False (F):

(i) All productive soils are fertile but all fertile soils are not productive.
(ii) The ability of a soil to meet the plant and ecosystem requirements for water, aeration and strength over time is called soil productivity.
(iii) The root zone and rhizosphere are two different terms.
(iv) The root zone is also called rhizosphere.
(v) The zone of rhizosphere having a high concentration of microorganisms is called rhizoplane.
(vi) Root microbiome has high contraction of soil biota.
(vii) The microbial abundance and diversity in the root microbiome are strongly influenced by the root activities.
(viii) Root exudates are toxic to soil microflora.
(ix) The zone of influence of plant roots varies generally between 1 and 100 mm.

(x) The root exudates in the rhizosphere influence acquisition of plant nutrients, especially Fe and P.
(xi) The root exudates are plant-species specific.
(xii) The population of useful bacteria increases in the root microbiome under water stress conditions.
(xiii) Plants can shape their rhizosphere microbiome to their own advantage.
(xiv) The characteristics of rhizosphere are independent of plant species.
(xv) Soil physical environment is highly complex and heterogeneous at spatial and temporal scales.
(xvi) Soil physical productivity is difficult to assess based on simple field observations.
(xvii) Pedotransfer function relates to complex soil physical processes which are difficult to measure.
(xviii) Weedy vegetation in the field is an indication of poor soil structure.
(xix) Restricted root growth is an indication of compact soil.
(xx) The LLWR is a valuable single value soil physical index.
(xxi) Any event, natural or anthropogenic, that causes a significant change from the normal pattern or functioning of an ecosystem is known as soil resilience.
(xxii) The ability to resist displacement from the antecedent state or the capacity of a soil to continue to function without change throughout the disturbance is called soil resilience.
(xxiii) Soil elasticity refers to the rate of recovery of soil after disturbance.
(xxiv) The soil resilience index is based on the soil quality index.
(xxv) The clay content is an important factor affecting soil resilience.
(xxvi) Green light is most effective for plant growth.
(xxvii) Red light is injurious to plant growth.
(xxviii) By virtue of their plasticity plants can adapt themselves to varying environmental conditions.
(xxix) Mesophytes are the plants which are adapted to drought conditions.
(xxx) Most of the seeds fail to germinate at temperatures >45 °C.

[**Key:** (i) T (ii) F (iii) T (iv) F (v) F (vi) T (vii) T (viii) F (ix) F (x) T (xi) T (xii) T (xiii) T (xiv) F (xv) T (xvi) F (xvii) F (xviii) F (xix) T (xx) T (xxi) F (xxii) F (xxiii) T (xxiv) T (xxv) T (xxvi) F (xxvii) F (xxviii) T (xxix) F (xxx) T].

1.9.5 Multiple Choice Questions:

(i) Plant growth occurs in

(a) Meristematic tissues
(b) Vascular tissues
(c) Epidermal tissues
(d) Parenchymatic tissues

(ii) Vegetative growth phase of a plant refers to

(a) Germination and emergence
(b) Emergence and flowering
(c) Between germination and emergence
(d) Between germination and flowering

(iii) Allelopathy refers to

(a) Biochemical catalytic effect between two living organisms
(b) Biochemical interaction between two living organisms
(c) Biochemical inhibition of one organism by the other
(d) Symbiotic interaction between mutually beneficial species

(iv) Altitude influences plant growth through temperature which decreases by

(a) 1 °C for every 100 m rise in altitude in dry air
(b) 1 °C for every 100 m rise in altitude in moist air
(c) 10 °C for every 100 m rise in altitude in dry air
(d) 10 °C for every 100 m rise in altitude in moist air

(v) Epiphytes are plant species that grow

(a) On surface of rocks
(b) On surface of other plants
(c) In soil
(d) In water

(vi) Biological properties refer to

(a) Soil flora
(b) Soil fauna
(c) Soil flora and fauna
(d) Soil flora, fauna and organic matter

(vii) To support metabolic processes within the seed embryo for germination the seeds must contain

(a) Around 10–25% water content
(b) Around 30–45% water content
(c) Around 50–65% water content
(d) Around 75–95% water content

(viii) Which of the following lights is most effective in plant growth

(a) Green light
(b) Red light
(c) Blue light
(d) White light

(ix) Flowering in plants is encouraged in

(a) Green light of solar radiation
(b) Red light of solar radiation
(c) Yellow light of solar radiation
(d) White light of solar radiation

(x) The *short-day plants* flower only when the day length is

(a) <3 h
(b) <6 h
(c) <9 h
(d) <12 h

(xi) The *long-day plants* flower only when day length is

(a) >3 h
(b) >6 h
(c) >9 h
(d) >12 h

(xii) The temperature range under which plants thrive and produce best is known as

(a) Optimum range
(b) Growth range
(c) Survival range
(d) Critical range

(xiii) Under excess water conditions, seeds may fail to germinate because of

(a) Water toxicity
(b) Oxygen deficiency
(c) CO_2 toxicity
(d) Low soil redox potential

(xiv) Rhizosphere refers to

(a) The root zone depth
(b) The root microbiome
(c) The soil in direct proximity of plant roots
(d) The top 15 cm soil layer

(xv) The ability of plant roots to adapt to variable environmental conditions is known as

(a) Root elasticity
(b) Root plasticity
(c) Geotropism
(d) Gravitropism

(xvi) The ability of a soil to meet the plant and ecosystem requirements for water, aeration and strength over time, and to resist and recover from processes that might diminish that property is known as

(a) Soil resilience
(b) Soil health
(c) Soil physical health
(d) Soil productivity

(xvii) The micro-ecological zone in direct proximity of plant roots is known as

(a) Root zone
(b) Rhizosphere
(c) Rhizoplane
(d) Root microbiome

(xviii) The predictive functions for some soil attributes using available minimum data set without actually measuring the specific soil attributes are known as

(a) Pedotransfer functions
(b) Pedological functions
(c) Soil health functions
(d) Soil physical indices

(xix) The upper limit of LLWR index is the water content at which

(a) Air-filled porosity is 0.1%
(b) Air-filled porosity is 1.0%
(c) Air-filled porosity is 10%
(d) Air-filled porosity is 100%

(xx) The lower limit of LLWR index is the water content at which

(a) Soil penetration resistance is 0.5 MPa
(b) Soil penetration resistance is 1.0 MPa
(c) Soil penetration resistance is 1.5 MPa
(d) Soil penetration resistance is 2.0 MPa

(xxi) The value of 'tilth index' developed by Singh et al. (1992) for poor and good soil physical health varies between

(a) 0 and 0.1
(b) 0 and 1.0
(c) 0 and 10
(d) 0.1 and 1.0

(xxii) According to 'soil physical quality index' developed by Gupta and Abrol (1993) a soil has poor physical health if the index value is less than

(a) 0.25
(b) 0.50
(c) 0.75
(d) 1.00.

(xxiii) Which one of the following factors reduces soil resilience

(a) Deep soil
(b) High sand content
(c) High clay content
(d) High organic matter content

(xxiv) One of the following characteristics do not apply to highly resilient soils:

(a) High buffering capacity
(b) High rates of recovery
(c) Large amplitudes
(d) No change with change in land use

[**Key:** i. a, ii. d, iii. c, iv. a, v. b, vi. d, vii. d, viii. b, ix. b, x. d, xi. d, xii. a, xiii. b, xiv. c, xv. b, xvi. c, xvii. b, xviii. a, xix. c, xx. d, xxi. b, xxii. a, xxiii. b, xxiv. d]

References

Bais HP, Park SW, Weir TL, Callaway RM, Vivanco JM (2004) How plants communicate using the underground information superhighway. Trends Plant Sci 9(1):26–32

Bakker PAHM, Berendsen RL, Doornbos RF, Wintermans PCA, Pieterse CMJ (2013) The rhizosphere revisited: root microbiomics. Front Plant Sci 4:7

Berendsen RL, Pieterse CMJ, Bakker PAHM (2012) The rhizosphere microbiome and plant health. Trends Plant Sci 17:478–486

Biswas S (2016) Soil resilience–a new approach to recover degraded soil. Biotech Articles. www.biotecharticles.com

Biswas S, Hazra GC, Purakayastha TJ, Saha N, Mitran T, Roy SS, Basak N, Mandal B (2017) Establishment of critical limits of indicators and indices of soil quality in rice-rice cropping systems under different soil orders. Geoderma 292:34–48

Bouma J (1989) Using soil survey data for quantitative land evaluation. Adv Soil Sci 9:177–213

da Silva AP, Kay BD (1967a) Estimating least limiting water range of soils from properties and management. Soil Sci Soc Am J 61:877–883

da Silva AP, Kay BD (1967b) Effect of soil water content on the variation in the least limiting water range of soils. Soil Sci Soc Am J 61:884–888

da Silva AP, Kay BD, Perfect E (1994) Characterization of least limiting water range of soils. Soil Sci Soc Am J 58:1775–1781

de Andrade BJ, Anghinoni I, de Moraes MT, Fink JR (2017) Resilience of soils with different texture, mineralogy and organic matter under long-term conservation systems. Soil Tillage Res 174:104–112

Dennis PG, Miller AJ, Hirsch PR (2010) Are root exudates more important than other sources of rhizodeposits in structuring rhizosphere bacterial communities? FEMS Microbiol Ecol 72(3):313–327

Dexter AR (2004) Soil physical quality, part I: theory, effects of soil texture, density, and organic matter, and effects on root growth. Geoderma 120:201–214

Garcia-Palacio P, Maestre FT, Bardgett RD, de Kroon H (2012) Plant responses to soil heterogeneity and global environmental change. J Ecol 100(6):1303–1314

Gupta RP, Abrol IP (1993) A study of some tillage practices for sustainable crop production in India. Soil Tillage Res 27:253–272

Hartmann A, Rothballer M, Schmid M (2008) Lorenz Hiltner, a pioneer in rhizosphere microbial ecology and soil bacteriology research. Plant Soil 312:7–14

Hiltner L (1904) Uber neuere Erfahrungen und Probleme auf dem Gebiete der Bodenbakteriologie unter besonderden berucksichtigung und Brache. Arb Dtsch Landwirtsch, Gesellschaft 98:59–78

Jones DL, Nguyen C, Finlay RD (2009) Carbon flow in the rhizosphere: carbon trading at the soil–root interface. Plant Soil 321:5–33

Lal R (1997) Degradation and resilience of soils. Phil Trans R Soc London B 352:997–1010

Lemon ER, Erickson AE (1952) The measurement of oxygen diffusion in the soil with a platinum electrode. Soil Sci Soc Am Proc 16:160–163

Letey J (1985) Relationship between soil physical properties and crop production. Adv Soil Sci 1:277–293

Lynch J (1995) Root architecture and plant productivity. Plant Physiol 109:7–13

Manna MC, Sharma KL (2009) Soil and water quality. In: Goswami NN, Rattan RK, Dev G, Narayanasamy G, Das DK, Sanyal SK, Pal DK, Rao DLN (eds) Fundamentals of soil science. Indian Society of Soil Science, New Delhi, pp 641–668

Marasco R, Rolli E, Ettoumi B, Vigani G, Mapelli F, Borin S, Abou-Hadid AF, El-Behairy UA, Sorlini C, Cherif A, Zocchi G, Daffonchio D (2012) A drought resistance-promoting microbiome is selected by root system under desert farming. PLoS One 7(10):e48479. https://doi.org/10.1371/journal.pone.0048479

McNear DH Jr (2013) The rhizosphere–roots, soil and everything in between. Nat Educ Know 4(3):1

Miller MB, Bassler BL (2001) Quorum sensing in bacteria. Annu Rev Microbiol 55:165–199

Mohanty M, Painuli DK, Misra AK, Ghosh PK (2006) Soil quality effects of tillage and residue under rice-wheat cropping on a vertisol in India. Soil Tillage Res 92:243–250

Mostafa E, Mehdi E, Baghernejad M, Fathi H, Saffari M (2008) Effect of land use change on selected soil physical and chemical properties in north highlands of Iran. J Appl Sci 8:496–502

Newman EI (1985) Ecological interactions in soil: plants, microbes and animals. In: Fitter AH, Atkinson D, Read DJ, Usher MBA, Fitter H (eds) The rhizosphere: carbon sources and microbial populations. Blackwell Scientific Publications, Oxford, pp 107–121

Rich SM, Watt M (2013) Soil conditions and cereal root system architecture: review and considerations for linking Darwin and weaver. J Exp Bot 64(5):1193–1208

Rickman RW, Letey J, Stolzy LH (1966) Plant response to oxygen supply and physical resistance in root environment. Soil Sci Soc Am Proc 30:304–307

Rovira AD (1969) Plant root exudates. Bot Rev 35:17–34

Sharma PK, Bhushan L (2001) Physical characterization of a soil amended with organic residue in a rice-wheat cropping system using a single value soil physical index. Soil Tillage Res 60:143–152

Singh KK, Colvin TS, Erbach DC, Mughal AQ (1992) Tilth index: an approach to quantifying soil tilth. Tans Am Soc Agril Biol Eng (ASAE) 35(6):1777–1785

Topp GC, Galganov YY, Wires KC, Culley JLV (1994) Non-limiting water range (NLWR): an approach for assessing soil structure. Soil quality evaluation program, Technical Report No. 2. Research Centre, Agriculture and Agri-Food Canada, Central Experimental Farm, Ottawa

Verma S, Sharma PK (2008) Long-term effects of organics, fertilizers and cropping systems on soil physical productivity evaluated using a single value index (NLWR). Soil Tillage Res 98:1–10

Vorholt JA (2012) Microbial life in the phyllosphere. Nat Rev Microbiol 10:828–840

Chapter 2
Soil Water and Plant Growth

Water is the key input in agriculture. Plants require water continuously from germination until maturity and that too in huge amounts. All the growth processes of plants including nutrient absorption, translocation, photosynthesis, respiration, transpiration, cell division, and several metabolic reactions within plant system and root zone are affected by water. Water helps in the maintenance of the plant's physical form by maintaining turgor pressure in plant cells. For optimum growth and yield of plants, soil water content must be in optimum range. The excess or deficit water conditions adversely affect plant growth processes. The soil water availability to plants varies widely with soil, plant, and environmental conditions. Plants differ significantly in their water requirements. The water-use efficiency by plants is linked with soil water regimes and soil and crop management skills. This chapter discusses the role of water in plants, concept of water availability to plants, crop water requirements, and water management practices under different water regimes.

2.1 The Role of Water in Plants

The water has significance for plants, both ecologically and physiologically:

(a) *Ecological significance*: The distribution of vegetation over the surface of the earth is largely controlled by the availability of water. For example, flush green vegetation (forest) grows in heavy-rainfall areas, while leafless cacti and other stem-succulent plants having erect growth dominate the arid and semiarid regions.

P. K. Sharma, S. Kumar, *Soil Physical Environment and Plant Growth*,
https://doi.org/10.1007/978-3-031-28057-3_2

(b) *Physiological significance*:

- It is a principal constituent of the growing plant. Around 80–95% plant tissue mass is composed of water. Water content in mature woody plant tissues ranges from 45 to 50%, while in herbaceous plants, it ranges from 70 to 95%.
- Water acts as a nutrient for plants. Most of the times, crop yield is determined by the availability of water rather than the deficiency of other plant nutrients.
- Water is a very good solvent. It dissolves gases, minerals (plant nutrients), and other solutes in soil. The plants absorb dissolved plant nutrients through soil solution. It acts as a carrier of food nutrients.
- Water acts as a reagent in many important processes, such as photosynthesis and hydrolysis of starch and sugar.
- Water maintains turgidity in plants. It imparts rigidity and mechanical stability in herbaceous plants, besides being essential for many physiological processes, such as cell enlargement and growth; opening of the stomata; movement of leaves, flowers, and petals; gas exchange in the leaves; and transport of water and sugars within plant tissues.
- Water regulates the temperature of plant and soil.
- Water impacts soil-forming processes and weathering of parent material into soil, providing a specific medium for plant growth.
- Almost all chemical and biological activities of soil are dependent on soil water content, which ultimately influences plant growth. Microorganisms require water for their metabolic activities, which in turn influence plant growth.

2.1.1 Impact of Waterlogging on Plant Growth

Waterlogging is a significant abiotic stress to crop production in high-rainfall areas and/or in low-permeable poorly drained soils. Waterlogging causes a series of physical, chemical, electrochemical, and biological modifications in soils, which lower the ability of soil to provide optimum medium for plant growth. It also affects a number of biological and chemical processes within the plant system. Because of such changes in soil and plant system, waterlogging may cause yield losses by 20–50% or even more. Soil saturation for 24 h may reduce growth and yield of most of the upland crops by 50%. Important changes in plant characteristics due to waterlogging/soil saturation are summarized in Table 2.1 (Manik et al. 2019).

Excess water *per se* does not react chemically with the plant and hence does not show direct phytotoxicity. It impacts crop plants primarily by causing oxygen deficiency (i.e., anoxia) in soil. Plants need oxygen for cell division, growth, and uptake and transport of nutrients. Oxygen supply in saturated/waterlogged soils cannot match with the oxygen requirements of plants because of lowered oxygen diffusion. Oxygen diffusion in waterlogged soils is 320K times lower than in non-submerged soils (Colmer and Greenway 2011). Generally, the oxygen level in a saturated soil declines to critical levels for plant growth within 48–96 h. The oxygen

Table 2.1 Changes in plant characteristics due to waterlogging/soil saturation

Plant part	Expected changes
Leaf	• Reduced photosynthetic activity • Reduced stomatal conductance
Stem	• Enhanced starch degradation, decreased soluble carbohydrates and ATP in cotyledons and hypocotyls • Reduced biosynthesis and enhanced degradation of chlorophyll
Roots	• Reduced root hydraulic conductivity • Decreased biological N fixation • Restricted root growth

Source: Adapted from Manik et al. (2019)

Table 2.2 Excess water tolerance of some crops

Crops	Excess water tolerance in ascending order
Cereals	Oats > wheat > barley
Pulses	Faba beans > soybeans > field beans > peas
Oilseeds	Canola (mustard) > sunflower > flax
Grass forage	Reed canary grass > timothy > orchard grass = perennial ryegrass
Legume forages	Bird's-foot trefoil/red clover > sweet clover > alfalfa

Source: Rajanna et al. (2018)

depletes faster at higher temperatures and in OM-rich soils having higher biological activity. Soils rich in actively metabolizing organic matter develop anoxia at a comparatively much faster rate. Plant tissues under anaerobic conditions use alternate metabolic pathways, which lead to the production of by-products, some of which are phytotoxic at relatively higher concentrations.

Plants usually suffer from N and P deficiencies in waterlogged soils. Waterlogging enhances nitrate leaching, increases denitrification, and reduces N mineralization. The denitrification may cause 1–5% N/day loss depending on the nitrate concentration, temperature of soil, and duration of saturation/waterlogging.

Waterlogging produces different types of phytotoxins in the root zone, which adversely affect root growth in several ways. Root growth is reduced, root rots are enhanced, and plants become sensitive to hot temperatures. They display N and P deficiencies, especially in the later part of their growth stage. Even if these visible symptoms do not appear, the crops invariably suffer from yield losses. Yield losses to the tune of 20–50% due to waterlogging have been reported in cereals, pulses, and oilseed crops (Manik et al. 2019).

Although excess water conditions harm almost all the arable crops, yet the degree of impact varies with the crop species. Some crops are more susceptible to waterlogging/soil saturation than others. Relative susceptibility of some crops to excess water is shown in Table 2.2.

2.1.2 Impact of Water Stress on Plant Growth

Water deficit adversely affects plant growth at physiological, morphological, and molecular levels. In general, there is reduction in synthetic processes and activation of degradation processes in plant system. Mild-to-moderate water stress reduces plant water potential and turgor to a level to interfere with the normal functions in plant system. Excessive water stress may lead to death of plant due to desiccation. Plant responses to water deficit largely depend on the degree and duration of water stress, time of occurrence of stress in relation to plant growth stage, plant species, and nature of plant produce.

2.1.2.1 Morphological Effects of Water Stress

Morphologically, plant processes such as root development, tiller formation, branching, flowering, seed formation, and seed development are adversely affected by water stress. There is reduction in plant height, leaf area index, and grain yield. There are increased incidences of early senescence in plants due to accumulation of abscisic acid. Reduction in the diameter of beetroot and onion bulb, intermodal length of sugarcane, leaf area per plant in tobacco, flowering and fruiting in most plants, incomplete filling of grains in cereals, fruit drop, and some such effects on many other crops are noticed under water-deficit conditions. The protein content of wheat grains and nicotine content of tobacco leaves increase with an increase in water stress. Vegetables and fodder crops are very sensitive to water stress. Even a mild water stress may significantly reduce their yield. Severe water stress for a longer duration significantly lowers the quality of vegetables, fodder, and fruits.

Certain plant growth stages, especially at which the cell division and differentiation are significant, are more sensitive to water stress. For example, a water deficit during crown root initiation stage in wheat, spike development stage in cereals, and branching, flowering, or seed development stages in legumes has more adverse effect on plant growth and yield.

2.1.2.2 Physiological Effects of Water Stress

Physiologically, the soil water stress reduces the soil water potential and the ability of plants to absorb water from soil. Water deficit within plant tissues reduces the rate of cell expansion in growing tissues. Cell growth and cell wall synthesis are reduced. The rate of formation of photosynthetic leaf area is reduced. It results in the reduction in the flow of assimilates to the meristematic and growing tissues of the plant, i.e., leaves and roots; leaves are affected more than the roots (Munns and Sharp 1993). There is loss of turgidity of leaves and closure of stomata. It decreases the photosynthetic assimilation of CO_2 in the leaves. Simultaneously, CO_2 metabolism is inhibited. Consequently, there is reduction in photosynthetic rate. There is reduction in nitrogen metabolism. There is reduction in carbohydrate assimilation, protein synthesis, and nitrate reductive activity.

2.1.2.3 Degree of Water Stress and Plant Growth

The level of water stress may be classified into three categories (Hsiao 1973):

(i) *Mild water stress*: The relative water content (RWC) of the plant drops by 8–10% of the value in a well-watered plant under conditions of low atmospheric evaporativity. It corresponds to a drop in plant water potential by −5 to −6 bar. Plants exhibit signs of wilting during the hottest part of the day only.

(ii) *Moderate water stress*: The RWC of the plant drops by 10–20% of the value in a well-watered plant under conditions of low atmospheric evaporativity. It corresponds to a drop in plant water potential by −12 to −15 bar. Plants show wilting for a considerable period during the daytime, and wilting recovers during the night.

(iii) *Severe water stress*: The RWC of the plant drops by >20% of the value in a well-watered plant under conditions of low atmospheric evaporativity. It corresponds to a drop in plant water potential by >−15 bar. Plants remain wilted continuously and do not properly recover at night, causing permanent leaf burning and ultimately death through desiccation.

The degree of impact of water deficit on plant growth depends on the severity and duration of water stress. If the duration of stress is brief, it may not cause a perceptible damage to certain types of crops such as grain crops. It is because these crops are able to compensate the damage by the subsequent development under no-stress condition. Water stress for a longer duration, however, damages plant growth and yield.

2.2 Soil-Plant-Water-Atmosphere Relations

Soil-plant-water-atmosphere relations relate to the physical properties of soil, plants, and atmosphere that influence retention, transport, and use of water within soil and plant systems. The concept connects soil, plant, and atmospheric water in a continuum, wherein water flows from soil to plants, through plants, and from plants to the atmosphere in response to water potential gradient.

Plants continuously absorb water from soil to meet their metabolic processes. The water absorption is much in excess of their metabolic requirement. Only about 0.5–1.0% water absorbed by plants is utilized for metabolic processes; the rest is lost to the atmosphere in the form of vapors, a process called *transpiration*. The transpiration is dictated by the atmospheric conditions, called the *atmospheric evaporativity*. The supply of water to plants is controlled by the soil properties. The water absorption and translocation by plants are, thus, decided not only by their physiology, but also by the soil hydraulic properties and atmospheric conditions. It means, therefore, that in order to understand water dynamics within the plant system, study of soil, plant, and atmosphere in a continuum is important.

2.3 Soil-Plant-Atmosphere Continuum (SPAC)

The water absorption by plants from soil is a dynamic process in which soil, plant, and atmosphere are physically integrated into one system. Plant water absorption involves soil-water status; plant characteristics, viz. root architecture, shoot growth, and canopy characteristics; and atmospheric evaporativity. John Philip (Philip 1966) was the first to formalize the concept and use the phrase "*soil-plant-atmosphere continuum*" (SPAC). The SPAC integrates soil, plant, and the surrounding atmosphere into a dynamic system in which the various transport processes involving energy and matter occur simultaneously. According to SPAC, water and energy transfer occurs from soil through plants to the atmosphere.

The SPAC may be defined as "the pathway for water moving from soil through plant to the atmosphere along an interconnected film of liquid water." The water in the SPAC flows in response to energy differences. The water potential gradient is the driving force. Water flows from a region of higher potential to the region of lower potential energy. The introduction of this concept has a practical advantage. It makes the understanding of the flux processes easier as it avoids the use of different terminologies to express water potential in soil (e.g., water suction or water tension), plant (e.g., diffusion pressure deficit), and atmosphere (e.g., vapor pressure or relative humidity).

As per the hydrological cycle, water molecules present in clouds as vapors condense and fall on the ground as rain. The water falling on the ground percolates into soil, and a part of it is temporarily retained in soil profile. During the process, some of it may be absorbed by the plant roots and lost into the atmosphere as vapors through the transpiration process. The return path of water from soil to atmosphere shows connectedness of soil, plant, and atmosphere. This forms the basis of concept of SPAC.

Two things are important to remember to understand the concept of SPAC. One, plants have a continuous demand for water from germination until maturity. Water is needed for various metabolic processes, transport of nutrients and metabolites within the plant, and maintenance of the plant's turgor pressure. Two, the concept of potential energy, as is used in soil, is also applicable to plant-water system. Water, as in soils, moves through plants along potential gradients. The flow path of water includes movement of water from soil to the plant roots in liquid and/or vapor form, absorption into roots to the vascular tubes of the xylem tissues in the stem, transfer through the xylem to the leaves, evaporation in the intercellular spaces within the leaves, and vapor diffusion through the sub-stomatal cavities and stomatal openings to the atmosphere. All through the way, water movement is passive caused by a potential gradient. The water potential is relatively high in soil and decreases progressively through plants and in atmosphere. It creates potential gradient across soil-plant-atmosphere continuum.

The variation in water potential at different boundaries in SPAC is shown in Fig. 2.1. It may be seen that the total potential difference between soil and atmosphere may amount to hundreds of bars and may cross 1000 bars in arid climates.

The main components of leaf water potential are the osmotic potential and the turgor potential. Since the water potential in the atmosphere is much negative

Fig. 2.1 Schematic sketch showing variation in water potential across SPAC

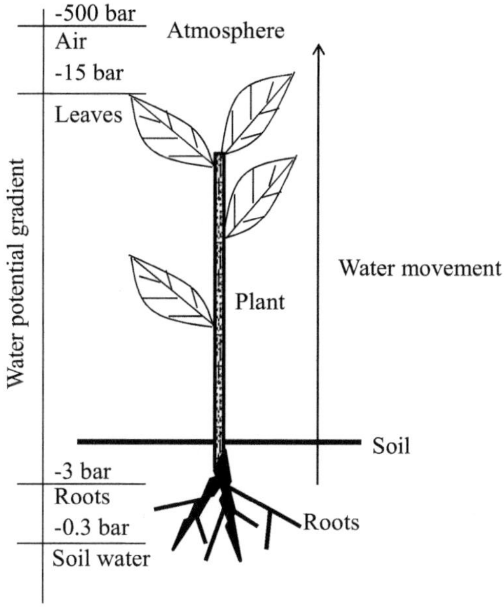

compared to the water potential in leaves, plants continuously keep on transpiring water into the atmosphere. The water content in leaves will remain unchanged as long as water flux from roots to leaves equals the rate of water loss through transpiration. Net removal of water from the leaf cells causes reduction in both the osmotic potential and the turgor potential, and the plants start showing the signs of wilting. For maintaining plant turgor, therefore, the water absorption by roots must equal the transpiration loss.

2.4 Nutrient-Water Interactions and Plant Growth

Nutrients and water are the two key inputs in crop production. Water is the lifeline of plants, and nutrients account for at least 35% of plant growth and yield. Both these factors have synergistic effects on nutrient and water-use efficiencies (Fig. 2.2). Crop response to one factor changes significantly by the other factor.

Nutrient applications increase overall growth and development of crops, affecting their ability to better utilize the nutrients and the water stored in the soil profile. Water supply affects various soil biological, chemical, and physical processes, thereby impacting nutrient availability to plants. Well-developed root system allows water uptake from deeper soil layers, particularly during drought spells. In one study, application of nitrogen to mustard crop resulted in 47% increase in root length index (RLI) and about 17% lower moisture storage (Yadav and Yadav 2008). Increases in crop yields and water-use efficiencies of different crops under rainfed situations with the applications of nutrients (NPK and FYM) have been revealed by several field investigations (Tandon 1987; Hati et al. 2006; Sun et al. 2009; Singh et al. 2015).

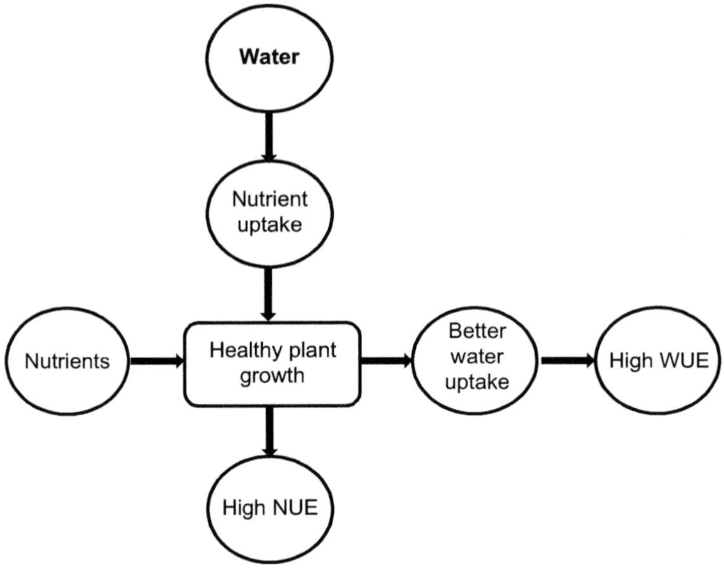

Fig. 2.2 Schematic diagram showing nutrient-water interaction effects on nutrient-use efficiency (NUE) and water-use efficiency (WUE) of crops

The entire process of nutrient-water movement within the soil system and uptake by the plants is highly dynamic and interactive. Water movement occurs due to potential gradients and is strongly influenced by ion concentrations in it. Simultaneous movement of water and solutes involves electrokinetic phenomena. All nutrient-related processes in soil, viz. *transformation* (mineralization, fixation, volatilization, etc.) and *translocation* (movement, leaching, uptake by plants, etc.), are highly dynamic and are mediated by soil water regimes. It suggests the indispensable interaction of the two factors in relation to nutrients' availability to plants. The quality of irrigation water is equally important in deciding nutrient and water-use efficiency of crops.

2.4.1 Nutrient Transformations and Soil Water Regime

The nutrient transformation processes decide the pool of plant available nutrients in soil. Optimum moisture availability in the soil plays a critical role in governing the population and activity of microbes responsible for transformation of nutrients. Highest microbial activity is observed at or near field capacity, with soil temperatures around 25–30 °C. As soil dries, microbial activity decreases with the resultant decline in organic matter breakdown and release of nutrients.

Mineralization is a major process in the soil that caters to the nutrient needs of the plants and is dependent on soil water regimes. N mineralization rate is negligible at wilting point and rises to a maximum at about field capacity (Gregory et al. 1997). Hu et al. (2014) observed highest net N mineralization rate at 85% field water content

(FWC). Nitrogen mineralization is highest at and around field capacity and negligible at wilting point. Irrigation or rainfall within about 36 h of urea application lowers N volatilization losses by moving surface-applied N into soil to undergo reaction. Waterlogging may increase oxygen stress and NO_3 losses through denitrification.

2.4.2 *Nutrient Translocation and Soil Water Regime*

The translocation of nutrients within soil profile has a significant effect on nutrient availability to plants. Plant roots access nutrients in soil in two ways: (i) root interception and (ii) movement of nutrients to plant roots. Plant roots grow to access less than 3% of the available nutrients in soil. Major part of nutrients' availability is accomplished through nutrient movement towards roots. Water acts as a vehicle for nutrients to move in soils to reach plant roots. Nutrients move in soil through *mass flow* and *diffusion*. Root interception is primarily responsible for the uptake of Mg and Ca. Mass flow supplies most of the N, S, Ca, and Mg, and to some extent K, to plant roots. Most of the P and K reach plant roots through diffusion. Micronutrients generally reach plant roots through diffusion. The relative significance of root interception, mass flow, and diffusion in ion transport to plant roots in soils is shown in Table 2.3.

2.5 Water Extraction Pattern of Crops

The water extraction pattern refers to "the relative amounts of water extracted by plant roots from different depths within the root zone." The water extraction pattern depends on the root architecture. A positive correlation is observed between water extraction and root distribution pattern. Plants generally have a higher concentration of roots in the upper part of the root zone and near to their base. In a normal soil with good aeration and without restrictive layers, a greater portion of roots of most crop plants remains within 45–60 cm soil depth and most of the water needs of plants are met from this zone. As the available water from this zone decreases, plants extract more water from lower depths. Because there are few roots in lower layers, the water extraction from lower layers may not meet water requirement even if sufficient water is available there.

Table 2.3 Relative contribution of root interception, mass flow, and diffusion in nutrient uptake by maize roots (Havlin et al. 2005)

	Root interception	Mass flow	Diffusion
Nutrient	Percent supply		
N	1	99	0
P	2	4	94
K	2	20	78
Ca	12	88	0
Mg	27	73	0
S	4	94	2

In a uniform soil profile with adequate moisture, crop plants usually extract 40% moisture from the first quarter of the root zone, 30% from the second, 20% from the third, and only 10% from the last quarter.

Crops and crop cultivars differ significantly in their rates of water extraction. Dardanelli et al. (1997) observed variation in water extraction rates of maize, sunflower, peanut, soybean, and alfalfa between 0.110 kl/day (by sunflower) and 0.029 kl/day (by alfalfa) (kl = kiloliter).

2.6 Crop Water Requirement

The amount of water required to grow a normal crop giving full yield potential is known as crop water requirement. Water use is estimated as mm of water applied or received as rainfall, converted to m^3/ha (1 mm = 10 m^3/ha). Different components of crop water requirement include water used in metabolic processes of the plant, transpiration, evaporation, water required for special operations in the field (viz. land preparation, transplanting, leaching of salts below the crop root zone, frost control), and water lost during irrigation (conveyance and application). The following terms, depending on what components are considered, are frequently used in defining and determining crop water requirement:

2.6.1 Crop Water Use (CWU)

The crop water use refers to "the amount of water used by the crop in metabolic processes for building plant tissues and the water lost as evapotranspiration." Crop water use has three components: water used in metabolic processes of the plant, transpiration, and evaporation from the cropped field. Only about 0.5–1.0% water absorbed by the plant is used in metabolic processes, and the rest (99–99.5%) is transpired by the plant. Since the water used in metabolic processes compared to transpiration and evaporation is negligibly small, the crop water use practically equals transpiration and evaporation, which collectively is called *evapotranspiration* (ET). Out of the total ET, evaporation accounts for about 20–30% and plant transpiration for the remaining 70–80%.

2.6.2 Transpiration (T)

Transpiration refers to the "loss of water from the leaves, young shoots, and flowers in the form of water vapors." Transpiration may also be defined as "the evaporative loss of water from plant foliage." Transpiration through shoots and flowers is negligibly small compared to transpiration from leaves. The leaf surface area available for transpiration is indicated by the leaf area index. The *leaf area index* (LAI) may be defined as "the ratio of leaf surface area (one side only) to land surface area."

2.6.3 Evapotranspiration (ET)

The evapotranspiration (ET) is defined as "the amount of water transpired by plant foliage and evaporation from the soil surface where plant is growing." In other words, ET is "the evaporation from vegetation-covered ground." Since it is difficult to differentiate between transpiration and evaporation under field conditions, both the processes are clubbed and termed as evapotranspiration.

2.6.4 Consumptive Water Use (CU)

The consumptive water use refers to "the amount of water used in the metabolic process of the plant and transpiration." The amount of water used for metabolism processes is a very small fraction of total water absorbed by the plant, i.e., 0.5–1.0%. Around 99–99.5% of the absorbed water is lost through transpiration. If all the water absorbed by the plants were used in metabolic processes and in the synthesis of plant tissues, only 1 mm equivalent absorbed water would produce record yields.

2.6.5 Crop Water Requirement (CWR)

The crop water requirement (CWR) refers to "the amount of water used by the crop in metabolic processes for building plant tissues and the water lost as evapotranspiration and during irrigation." In simple words, CWR is the sum of crop water use (CU) and water lost during irrigation.

2.6.6 Net Water Requirement (NWR)

The net water requirement (NWR) refers to "the amount of water required to replenish evapotranspiration and deep percolation/seepage or root zone soil moisture deficit."

2.6.7 Gross Water Requirement (GWR)

The gross water requirement (GWR) refers to "the actual amount of water supplied to meet crop evapotranspiration and percolation/seepage observed under field conditions."

2.6.8 Total Crop Water Requirement (TWR)

The total water requirement of the crop refers to "the sum of crop water requirement (CWR) and water required for special operations such as land preparation, transplanting, leaching of salts below the crop root zone, and frost control." In simple words, the TWR refers to the total water required for raising the crop. Most of the times, however, the "water used in special operations" is also considered as a component of crop water requirement (CWR), and, thus, crop water requirement and total crop water requirement become synonymous terms, i.e.,

$$\text{TWR} = \text{CWR} = \text{CWU} + \text{Water lost during irrigation} + \text{Special water needs}$$

2.6.9 Factors Affecting Crop Water Requirement

The crop water requirement (CWR) depends on several factors including crop species, soil properties, irrigation method, cultural practices, and climatic conditions (temperature, solar radiation, humidity, wind, etc.). Under a given set of climate, soil, and management conditions, the variation in CWR among different crops is quite large. For example, the seasonal water requirement of corn, rice, wheat, and sugarcane may vary between 500 and 800 mm, 900 and 2500 mm, 450 and 650 mm, and 1500 and 2500 mm, respectively, depending on soil properties and climatic factors like sunshine, temperature, humidity, and wind speed. In terms of volume, 1 mm/ha is equal to 10,000 L of water per hectare. Crop water requirement of some crops is shown in Table 2.4.

2.7 Water-Use Efficiency (WUE)

The water-use efficiency (WUE) in agriculture in simple words means how efficiently the water is used for crop production. The objective is to maximize crop yield per unit of water use, i.e., *more crop per drop*, or minimize water use without compromising with crop yields, soil productivity, water resources, and environmental quality. The WUE may be addressed in two ways: (a) hydrological concept and (b) physiological concept.

2.7.1 Hydrological Concept of WUE

The hydrological concept of WUE is the domain of irrigation and water conservation engineers. This concept is used to describe how effectively water is delivered to the field to grow crops and to indicate the amount of water wasted. It does not take into

Table 2.4 Crop water requirement of different crops of middle Gujarat, India (Mehta and Pandey 2016)

Crops	Seasonal total (mm)
Kharif season	
Rice	729.3
Maize	445.4
Pearl millet	323.6
Green gram	324.6
Soybean	533.3
Groundnut	536.5
Cotton	848.0
Winter season	
Wheat	501.2
Maize	420.3
Chickpea	411.7
Mustard	469.1
Summer season	
Pearl millet	499.2
Green gram	476.5
Groundnut	849.0

consideration the biomass production or crop yield or economic gains from water use. The irrigation specialists use the term "*irrigation-use efficiency*" to describe this concept. Israelsen (1932) defined water (irrigation)-use efficiency as "the ratio of irrigation water transpired by the crops during their growth period to the water diverted (from the source) into the farm during the same period of time."

2.7.2 Physiological Concept of WUE

The physiological concept of WUE is used by the crop scientists. It is a plant-based concept relating biomass/grain yield production with the water use. In the beginning of the twentieth century, *transpiration ratio* was used as an index of crop water requirement. Transpiration ratio refers to "the quantity of water transpired by a crop to produce a unit amount of dry matter." The first study relating plant growth to transpiration appears to have been conducted by John Woodward (1665–1728) as reported by Stanhill (1986). Subsequently, significant work on these lines was conducted at Rothamsted Experimental Station (Lawes 1850).

Plants transpire huge amounts of water to produce unit quantity of biomass. For example, one hectare of corn may transpire between 28,000 and 37,000 L of water per day. The transpiration ratio for most of the field crops varies between 200 and 1000 kg water transpired/kg of dry matter produced (Martin et al. 1976). Field determinations being tedious, transpiration ratios for different crops have mostly been determined under controlled conditions. The transpiration ratio concept is often used in crop water modeling studies.

Briggs and Shantz (1913) first time used the term *water-use efficiency* (WUE) for the reciprocal of transpiration ratio. The plant WUE was defined as "the amount of biomass/grain yield produced per unit quantity of water transpired by the crop during the cropping season." Crop scientists now define WUE in different ways (Sinclair et al. 1984). The plant WUE refers to "the ratio of total biomass or grain production to evapotranspiration." The agronomic WUE refers to "the ratio of total biomass or grain production to the total water use on a daily or seasonal basis." The crop physiologists define WUE as "the amount of carbon assimilated as biomass or grain yield per unit quantity of crop water use."

$$Plant\ WUE = \text{Aboveground biomass or grain yield / evapotranspiration}$$

$$Agronomic\ WUE = \text{Aboveground biomass or grain yield / total crop water use}$$

Different components of total crop water use include volume of water lost by runoff from the field, *water* drained out of the root zone by deep percolation, *water* lost by evaporation during delivery and application of water to the field, water evaporated from the soil, water transpired by weeds, and water transpired by the crop. All these volumes pertain to the same unit area and time period. The agronomic water-use efficiency of some crops in India is given in Table 2.5.

2.8 Measurement of Water-Use Efficiency

Measurement of water-use efficiency (WUE) requires knowledge of two parameters, i.e., the total biomass or grain yield of the crop and the crop water use (i.e., evapotranspiration, ET). It is easy to determine crop biomass or grain yield. The real problem is precise measurement of ET. The ET can be measured directly or estimated from empirical and analytical equations using local climate data. It may be noted that most agronomic studies, however, do not determine ET but instead use the amount of water used to achieve the crop yield to calculate WUE. Hence, the published data on WUE may be carefully interpreted (Steduto 1966).

Table 2.5 Agronomic water-use efficiency (WUE) of some of the crops grown in India (Yadav et al. 2000)

Crops	No. of sites/observations	WUE range (kg m^{-3})	Average WUE (kg m^{-3})
Rice	6	0.30–0.54	0.45
Wheat	23	0.58–2.25	1.24
Maize (rabi)	10	0.49–1.63	0.91
Sorghum	7	0.56–1.43	0.88
Chickpea	8	0.40–4.02	1.60
Lentil	6	0.39–2.43	1.05
Green gram	4	0.37–0.50	0.44

2.8.1 Direct Measurement of Evapotranspiration (ET)

Direct measurement of ET is more accurate and reliable, but the methods are laborious, time consuming, and expensive. Two types of direct measurement methods are in common use: *water budget* and *water vapor transfer* methods.

The WUE of the individual leaves or the parts of leaves can be measured in the greenhouse or under field conditions by using portable *photosynthetic devices*. These devices determine photosynthetic rates and transpiration rates of the leaf. The WUE of leaves under laboratory conditions can also be determined by using *carbon isotope discrimination method* using mass spectrometer.

2.8.1.1 Water Budget Methods

In water budget methods, all water budget components are measured and the loss of water is determined as ET. The ET is calculated as the difference between input and output of water in a system; the system may be a small container (lysimeter) or a large catchment area. Following is the general water balance equation:

$$P = Q + ET + \Delta S \quad (2.1)$$

$$\text{or} \quad ET = P - Q - \Delta S \quad (2.2)$$

where P is the precipitation and irrigation (water input), Q the surface runoff or drainage (water output, i.e., loss), ET the evapotranspiration, and ΔS the change in soil water storage. If there is contribution from groundwater (as under shallow water table conditions), the component (D) is added to the input term in Eq. (2.1) and the equation becomes

$$P + D = Q + ET + \Delta S \quad (2.3)$$

$$\text{or} \quad ET = (P + D) - (Q + \Delta S) \quad (2.4)$$

Following are some of the reliable and time-tested water budget methods for determining ET:

(a) Use of water balance equation:

Different components of water budget Eq. (2.1), viz. P, Q, and ΔS, are determined or estimated in the field or catchment, and the difference between input and output gives ET during the period of measurements (Eq. 2.2).

(b) *Lysimeter*:

In this method, water budget components are measured using a *lysimeter* (non-weighing or weighing type). In case of *non-weighing* lysimeters, all components of water budget are measured, including rainfall and irrigation (water input), drainage (water output), and change in soil water storage, and ET is computed using water balance Eq. (2.1). In case of *weighing type* of lysimeters, ET is measured directly by measuring the change in weight of the lysimeter.

(c) *Soil moisture depletion method*:

In this method, components of water budget equation are actually measured *in situ*. Measured quantity of irrigation is applied (I), effective rainfall (the amount of rainfall actually received in the root zone) is determined (R), deep percolation or runoff is checked or otherwise measured (Q), and soil wetness in the root zone at the beginning (w_i) and at the end of the season (w_f) or at the time of applying irrigations, etc. is determined. The *ET* is then computed as follows:

$$ET = I + R + \left(w_{\mathrm{i}} - w_{\mathrm{f}}\right) \tag{2.5}$$

This method is very accurate but site specific. The information about ET at different crop stages depends on the frequency of measurement of soil wetness in the root zone. If soil wetness is measured only at the beginning and end of the experiment, it gives average *ET* value for the entire crop season.

(d) *Remote sensing*:

The ET on regional basis is measured by combining conventional meteorological measurements with remotely sensed data. *Remote sensing* refers to "observation and measurement of objects from a distance." The objects emit, transmit, or reflect some kind of energy, which is measured from a distance using specially designed sensors. Remotely sensed values are average values, while *in situ* measurements are usually point measurements. Compared to conventional methods, remote sensing has the advantage of collecting huge amount of data in a short time and that too at a low operational cost. Further, the meteorological parameter profiling is done in continuity with time, which is difficult with conventional methods. Although recent advances in remote sensing have further improved the estimation of ET, quantifying ET from mixed vegetation environments is still challenging because of the heterogeneity of plant species, canopy covers, and microclimates. Ground validation of remotely sensed average values with in situ measurements may also become difficult. Still remote sensing-based ET models can accurately estimate regional ET values.

2.8.1.2 Water Vapor Transfer Methods

In these methods, the flow of water vapor into the atmosphere (ET) is measured using meteorological sensors mounted above the ground surface. These sensors measure evaporation in terms of mass, or in the context of the surface energy balance as latent heat flux. The heat energy is transferred to the atmosphere with the water vapor in the form of latent heat.

The actual ET is calculated from the energy balance equation as follows:

$$\lambda ET = Rn - G - H \tag{2.6}$$

where λET is the latent heat flux (i.e., energy needed to change water from liquid to gaseous phase), Rn the net radiation, G the soil heat flux, and H the sensible heat flux. The Rn and G can be measured or estimated from climatic parameters. Measurement of H is complex, which requires data on temperature gradients above the soil surface. The components of the energy balance can be calculated using instruments like soil heat flux plates, radiation meters, or scintillometer. The energy available for actual ET can then be solved.

Eddy covariance is another technique of measuring the actual ET under field conditions. In this method, fast fluctuations of vertical wind speed are correlated with fast fluctuations in atmospheric water vapor density. This directly estimates the transfer of water vapor (ET) from the land surface to the atmosphere.

2.8.2 Indirect Measurement of Evapotranspiration (ET)

These methods estimate the actual crop ET (AET) from the reference evapotranspiration (ET_0) values using crop coefficients (k_c). Different crops exhibit different ET under the same weather conditions. The AET depends on crop characteristics (type of crop, crop stage, crop height, crop roughness, canopy cover and albedo, etc.), time of planting, soil characteristics, and weather parameters. The same factors influence crop coefficients (k_c).

Reference evapotranspiration (ET_0) represents the rate of evapotranspiration of green grass under ideal conditions, 8–15 cm tall, with extensive vegetative cover completely shading the ground. The ET_0 can be estimated by using *theoretical* (based on physical processes), *analytical* (based on energy or water budgets) or *empirical* (based on observational data) models. Some of the commonly used models for estimating ET_o are:

(a) Thornthwaite method (Thornthwaite 1948)
(b) Blaney-Criddle method (Blaney and Criddle 1950)
(c) Penman method (Penman 1948)
(d) Christiansen method (Christiansen 1968)

The most practical method for determining ET_0 is the pan evaporation method, using the Class A evaporation pan (circular) or the Colorado sunken pan (square). The Class A evaporation pan is more commonly used.

The crop factor (k_c) is determined experimentally for each crop and location and at different crop growth stages. The crop ET is generally low during early crop stage and maximum during active vegetative stage and then declines during ripening stage. Accordingly, the value of k_c during the initial stage for a well-watered crop with little vegetation may be around 0.35. As the vegetation develops, k_c increases to a maximum value and then declines as the crop matures and its moisture requirement diminishes. The value of k_c may vary between 0.2 and 1.3.

The k_c is generally determined during the following four growth stages:

(i) *Initial stage*: between emergence and early crop growth when soil cover is less than 10%
(ii) *Crop development stage*: between the end of initial stage and the stage when the crop canopy provides 70–80% soil cover
(iii) *Midseason stage*: between the end of crop development stage and the start of maturity stage (generally beyond flowering stage)
(iv) *Late-season stage*: between the end of the midseason stage and the full maturity or harvesting stage

The actual crop ET (AET) is then calculated from reference ET (ET_o) using the following equation:

$$\mathrm{AET} = kc \times \mathrm{ET}_o \tag{2.7}$$

The AET is lower than ET_0, maybe around 2/3rd of ET_0 value.

2.9 Increasing Water-Use Efficiency (WUE)

Conceptually, WUE can be increased by increasing crop productivity per unit of water use and/or decreasing evapotranspiration (ET) losses. Increase in crop yields without increasing ET and decrease in ET without affecting crop yields are possible at plant and field levels. Both the options are briefly discussed below:

2.9.1 Increasing Crop Yields Per Unit of ET

2.9.1.1 Growing Water-Efficient Crops

Selection of crops and crop cultivars must depend on their water requirement and suitability to the agroecological situation. Short-duration, deep-rooted, and drought-resistant crops are well suited to water-deficit and rainfed situations. Pulses and oilseeds (peas, lentil, mustard, sesame, etc.) may be preferred over cereals (maize) in water-deficit areas.

Sharma and Bhushan (1999) investigated under controlled conditions two rice cultivars for evapotranspiration (ET) rates in relation to four water regimes, viz. 3 cm continuous submergence, saturation, and −20 kPa and −50 kPa matric potential. They concluded that (i) potential ET of wetland rice cultivar under optimum water conditions was higher than upland rice cultivar; (ii) crop ET declined

Table 2.6 Seasonal evapotranspiration (ET), biomass production, and WUE of two rice cultivars under different water regimes (Sharma and Bhushan 1999)

Rice cultivar	Water regime	Seasonal ET (mm)	Biomass (g)	WUE (g biomass/mm water)
HPU 2216 (cultivar recommended for wetland conditions)	3-cm submergence	193.7 (1.00)	23.0 (1.00)	0.119 (1.00)
	Saturation	144.8 (0.75)	16.8 (0.73)	0.116 (0.97)
	−20 kPa matric suction	95.8 (0.49)	6.7 (0.29)	0.070 (0.59)
	−50 kPa matric suction	88.4 (0.46)	5.0 (0.22)	0.056 (0.47)
	LSD (0.05)	21.6	2.2	
Dhan 22 (cultivar recommended for upland conditions)	3 cm submergence	169.5 (1.00)	12.3 (1.00)	0.073 (1.00)
	Saturation	136.1 (0.80)	11.9 (0.97)	0.087 (1.19)
	−20 kPa matric suction	107.9 (0.64)	12.4 (1.01)	0.115 (1.57)
	−50 kPa matric suction	102.3 (0.60)	12.4 (1.01)	0.121 (1.66)
	LSD (0.05)	19.9	Ns	–

Note: Values in parentheses are ratios with values at 3 cm submergence

in both cultivars with the imposition of water stress, but the decline was more in wetland than in upland rice; and (iii) response of biomass production to decline in ET due to water stress was different in two rice cultivars; biomass production reduced drastically in wetland rice but remained unaffected in upland rice (Table 2.6). The seasonal ET of wetland rice at saturation and −20 and −50 kPa matric potential was, respectively, 75%, 49%, and 46% of potential ET, while that of upland rice was 80%, 64%, and 60% of potential ET, respectively. The biomass production of wetland rice at saturation and −20 kPa and −50 kPa matric potential was 73%, 29%, and 22% of that at 3 cm submergence, respectively, but it remained unchanged in upland rice. WUE declined with water stress in wetland rice but increased in upland rice.

2.9.1.2 Breeding High-Yielding Cultivars

Breeding crops and crop cultivars having traits of resistance towards abiotic (water, nutrient, salt, and temperature stresses) and biotic stresses (pests and diseases), developing water- and nutrient-efficient crops, changing harvest index and stay-green factor in crops, etc. increase crop yields without a concomitant increase in crop ET.

2.9.1.3 Reduction in Transpiration

The use of various antitranspirants to reduce transpiration received much attention during the 1960s and 1970s. Different materials have been tested as antitranspirants, such as phenyl mercuric acetate, atrazine, kaolin, or various long-chain alcohols, with a limited success; daily ET reduced by 5–10% only. The effect of antitranspirants persisted maximum up to 2 weeks. The long-chain alcohols blocked photosynthesis more than the transpiration because CO_2 molecules are 1.6 times bigger than water molecules. Different studies have yielded variable results. The technology has also not been found economically and practically feasible. It must be understood that reduction in transpiration may be translated into reduced crop yields. There is a trade-off between the two. Because of such limitations, use of antitranspirants has not become popular with the farming community for increasing WUE (Evans and Sadler 2008).

Regular weed control through mechanical or chemical means reduces transpiration from weeds, which is a nonproductive water loss. It makes more water available to crop plants. Weed control also reduces competition of weeds with the main crop for nutrients, light, water, space, etc. and helps in raising crop yield.

2.9.1.4 Reduction in Evaporation

Adoption of practices like crop residue management, mulching, growing cover crops, and narrow spacing decreases evaporation from soil surface. Reduction in evaporation diverts more water into transpiration. Crop yields are increased. It results in higher WUE. In a field study under temperate conditions, Kapur et al. (1978) concluded that pine needle mulch produced 34% higher grain yield and 39% higher WUE of rainfed wheat as compared to no-mulch treatment (Table 2.7).

2.9.1.5 Irrigation Management

Rainfed crops may experience short-term droughts even in areas receiving reasonably sufficient rainfall (for example, humid and temperate areas in the Mississippi River Delta and the southeast United States; high-rainfall areas in the

Table 2.7 Effect of pine needle mulch on grain yield and WUE of rainfed wheat under temperate conditions (Kapur et al. 1978)

	Grain yield (t/ha)		WUE (kg grains/mm-ha)	
Treatment	1975–1976	1976–1977	1975–1976	1976–1977
Mulch[a]	4.76	3.81	9.06	12.39
No mulch	3.61	2.64	6.87	8.40
LSD (0.05)	2.8	1.1	–	–

[a]Pine needle mulch @ 8 t/ha applied immediately after sowing of wheat

north-west Himalayan region, India) resulting in reduced yields. Same is true in partially irrigated areas. Introduction of supplemental irrigation in such areas, especially at critical growth stages, helps in stabilizing crop yields, without much affecting the ET.

Deficit irrigation scheduling also helps in improving WUE. Deficit irrigation means application of water to crops, which is less than the normal recommended schedule. Deficit irrigation is actually a trade-off between water use and grain yield. Applying irrigation only at the most critical crop growth stages saves water without significant decline in grain yield. The saved water may be diverted to high-value crops. It also saves on the cost of fuel, electricity, and labor. Deficit irrigation is currently being used in large areas in the Texas High Plains and in the Columbia Basin in Oregon and Washington on annual crops and on perennial crops (Evans and Sadler 2008).

Different crops respond much differently to deficit irrigation. For example, some potato varieties can withstand very little water stress without significant quality problems, whereas wine grapes (*Vitis vinifera*) can produce high-quality grapes with only about 50% of that of a fully irrigated vine (Evans 1999). Sharma (1989) obtained higher WUE but lower yields of transplanted rice under deficit irrigation conditions (i.e., −30 kPa matric potential) than under normal water conditions (i.e., continuous 5 cm flooding and irrigation at −10 kPa matric potential) (Table 2.8). WUE is generally higher under rainfed than under irrigated conditions, although the rainfed crop yields usually remain comparatively low.

Adoption of drip, trickle, or sprinkler irrigation and use of fertigation help in improving crop yields without much change in consumptive water use.

2.9.1.6 Improved Agronomic Practices

Agronomic practices such as conservation tillage, organic residue management, SOC buildup, water and nutrient management, crop geometry, planting densities, improved varieties, and disease and pest control significantly affect crop yields and

Table 2.8 Effect of irrigation management on rice grain yield and water-use efficiency (WUE) (Sharma 1989)

Irrigation level	Total water use[a] (mm)	Rice grain yield (kg/ha)	WUE (kg/ha-mm)
Continuous flooding	3640	5100	1.40
Flooding at −100 cm matric potential	2797	4700	1.68
Flooding at −300 cm matric potential	1870	3800	2.03
LSD (0.05)	–	795	–

Note: Flooding depth in each case was 5 cm; the data on water use, grain yield, and WUE are mean values of 2 years (1986 and 1987)

[a]Irrigation + rainfall

water use. Cropping strategies such as double-cropping, intercropping, relay cropping, and crop-specific rotations and sequences can take advantage of the lower water demand of certain crops and times or periods with higher rainfall to increase productivity. All such practices improve crop WUE.

2.10 Crop Water Productivity

The productivity in general refers to the input-output ratio. Crop yield per unit area (called crop productivity), crop yield per unit of ET (called WUE), and crop yield per unit of crop water use (called crop water productivity) are commonly used in water management studies. The *crop water productivity* is defined as "the ratio of grain yield (or biomass) to the total crop water use." Water-use efficiency (WUE) indicates how efficiently a crop utilizes water available in the root zone (ET) in producing biomass. It says nothing about the total amount of water used in raising a crop. The crop water use may be twice or even more as high as ET. The crop water productivity, on the other hand, indicates the total water required to raise the crop. Knowledge of crop water requirement is essential for irrigation projects' planning and development on a basin level.

Crop water productivity is significantly influenced by several factors like climatic parameters (CO_2 concentration, temperature, rainfall, humidity, atmospheric evaporativity, crop-growing season, etc.), crop species, soil fertility, fertilizer use, and irrigation efficiency. In a field study conducted in a sandy loam soil under rainfed conditions, Sharma and Kharwara (1990) found crop water productivity of wheat, barley, chickpea, and lentil (119 mm seasonal rainfall) as 10.38, 15.74, 5.36, and 6.26 kg/ha-mm, respectively. Seasonal rainfall (119 mm) plus profile-water depletion, instead of ET, was used as a denominator to compute crop water productivity.

Agronomists have always been trying to increase crop water productivity by adopting scientific agronomic practices of crop production. Earlier, the focus has been on increasing crop yields per unit area. But with the declining yield growth rates or even stagnation of crop yields per unit area (may be due to fatigued soils), the focus shifted to the improved water management practices with the objective of reducing total crop water requirement without affecting crop yields.

There are possibly three complementary and overlapping options to reduce/save water at the farm and field level and enhance crop water productivity: (i) increasing crop productivity per unit of water, (ii) spatially optimizing water applications and use, and (iii) improving management capacity of the growers. The first option considers agronomic alternatives that include planting drought-tolerant cultivars, reducing inputs such as fertilizers or water to decrease vegetative vigor (increased vegetative vigor increases rather than reducing crop water use), disease and pest control, etc. Option (ii) can be achieved through scientific irrigation scheduling, deficit irrigation, cultivation of crops suited to agroecology, and diverting irrigation water from less productive to more productive areas or uses. Option (iii) requires increasing growers' ability to optimize irrigation amounts in time and space, to adopt site-specific irrigation techniques, to enhance water delivery systems, and to use decision support tools and other advanced water management methodologies.

2.11 Water Productivity: A Neoclassical Concept of Water Management

The classical concepts of WUE and crop water productivity underestimate the irrigation efficiency. The term irrigation efficiency is frequently used by the irrigation engineers and water managers. *Irrigation efficiency* is defined as "the ratio of the amount of water consumed by the crop (ET) to the amount of water supplied through irrigation." The water engineers are usually concerned with (a) the assessment of the quantum of water loss during conveyance from the source to the field (i.e., conveyance efficiency, distribution efficiency, application efficiency, storage efficiency) and (b) identification of interventions to reduce water losses so as to improve the *irrigation efficiency*.

Classically, irrigation efficiency considers the water lost during conveyance (seepage, percolation, evaporation, runoff, etc.) and application to the field (nonuniformity of water application, deep percolation, etc.) as the *net* water loss from the system. Since this loss of water is a component of total crop water requirement and is used as a denominator for calculating crop water productivity, it underestimates irrigation efficiency. The low irrigation efficiency may lead to a perception that irrigation is a wasteful practice in crop production.

Water experts now realize that the "water losses," as defined by irrigation efficiency, may actually be captured and recycled for use elsewhere in the basin, for example, for recharging groundwater, leaching salts and phytotoxins from the root zone, and proving downstream sources of water for agriculture and other ecosystem services. Thus, the classical concept of crop water productivity and irrigation efficiency may be misleading. These concepts fail to assess the performance of water use in a large system, i.e., a basin or sub-basin system.

Jensen (1977) proposed revising classical "irrigation efficiency" (CE) to "net irrigation efficiency" (NE), also called "effective irrigation efficiency." He defined NE as

$$\mathrm{NE} = \mathrm{CE} + E_{\mathrm{r}}\left(1 - \mathrm{CE}\right) \tag{2.8}$$

where E_{r} is the percentage of (1 − CE) that is potentially available for recovery, reuse, or recycling somewhere in the hydrological system. The term (1 − CE) is called *classical inefficiency*.

Let us consider an irrigation system having 40% efficiency (i.e., CE = 40%). It means 40% of field-diverted water is used as crop ET and the rest 60% is lost during irrigation. If 70% of the so-called lost water is potentially available for reuse, then

$$NE = 0.40 + 0.70(1 - 0.40) = 0.40 + 0.42 = 0.82, \text{i.e.}, 82\%$$

Thus, with the same parameters, the net irrigation efficiency is more than twice as high as classical irrigation efficiency.

To address such ambiguities, the concept of *water productivity* (WP) was introduced (Kijne et al. 2003a, b; Seckler et al. 2003). *Water productivity* may be defined as "the ratio of the net benefits from varied outputs (viz. crops, forestry, fishery, livestock, and mixed farming systems) to the amount of water required to produce those benefits (Molden et al. 2007). WP is a measure of the economic, livelihood, or biophysical outputs derived from the use of a unit of water. WP considers multiple water use and reuse for a wide variety of outputs, viz. crop production, forestry, fishery, animal husbandry, and several other ecosystem services on a basin scale. Units of WP may be jobs/m^3, \$/m^3, or total biomass (kg/m^3). According to this perspective, the water losses occurring during irrigation may not be the *real* water losses to the system as a whole. Water lost during conveyance or application to the fields through seepage or percolation or surface runoff may be captured for reuse or recycled in other service sectors in the same basin. Only the water involved in non-beneficial uses, such as transpiration from weeds, evaporation from water surfaces, and water lost to permanent sinks (saline water), constitutes the real water losses. Improving WP seeks to *save water* and get the highest benefits from it. *Water saving* is defined as "the process of reducing non-beneficial water uses and making the water available for more productive uses."

The WP may be distinguished into physical and economical productivity. The *physical water productivity* refers to "the ratio of the mass of agricultural output to the amount of water used." It may be expressed as food production (kcal/m^3). The *economic water productivity* is defined as "the value derived per unit of water used." It may be expressed in terms of monetary value (\$/m^3).

Water productivity is a neoclassical concept of water management. It has given a new dimension to policy planning and water management. It is a holistic and integrated approach to assess the performance of water on a larger scale, i.e., basin or sub-basin scale. The classical irrigation efficiency (CE) operates on a smaller scale, i.e., at a field or farm level. As the scale of the system increases from field through farm to the basin level, CE decreases because of increase in water losses, while WP increases because of increased possibilities of water recycling and reuse. Studies conducted in the Nile irrigation system in Egypt found average CE to be around 50% while WP was around 87% (Molden et al. 1998). WP emphasizes on saving the productive water losses for reuse elsewhere in the same basin, maximizing benefits of captured or recycled water (reallocating water from lower to high-value uses), reducing non-beneficial water losses (evaporation and discharges to sinks), and reducing water pollution, waterlogging, and flood damages. The WP concept has become more relevant under the scenario of shrinking water resources and increasing water demands by increasing population and nonagricultural service sectors.

2.12 Water Management Under Different Water Regimes

Usually, three types of water regimes are encountered under field conditions and the objective of water management under these situations is situation specific:

(i) Excess water conditions: the objective is to improve *land productivity*, i.e., crop yield per unit area.
(ii) Irrigated conditions: the objective is to improve *crop water productivity*, i.e., *more crop per drop*.
(iii) Rainfed conditions: the objective is to improve *crop productivity*.

Each type of water regime has its own constraints and potentials and requires specific technological interventions for its management.

2.12.1 Water Management Under Excess Water Conditions

A wetland is an ecosystem that remains flooded or saturated with water permanently or seasonally. Waterlogging occurs due to high rainfall and/or poor drainage conditions. Saline soil conditions also favor waterlogging due to reduced percolation. The water in wetlands may be freshwater, brackish water, or saltwater. Global inland and coastal wetlands are estimated to be around over 1210 million ha, with 54% permanently inundated and 46% seasonally inundated (Ramsar Convention on Wetlands 2018). Around 16% of the soils in the United States, and 10% of the agricultural lands of Russia and irrigated areas of India, Pakistan, Bangladesh, and China are affected by waterlogging (FAO 2015). Global climate change causes waterlogging events to be more frequent, severe, and unpredictable. Some currently wet areas will become wetter, and prolonged waterlogging will also become more prevalent (IPCC 2014). It has been estimated that over the years, with increasing pressure on agriculture, around 50% of the world's wetlands will be lost (Verhoeven and Setter 2010). Several of the wetlands for millennia have also been drained and developed for agriculture.

The objective of water management under excess water conditions is to *maximize land productivity*, i.e., crop production per unit area. Wetlands are difficult to till. Heavy machines are difficult to operate due to soft consistency. Tilling wetlands may lead to puddling and destruction of soil structure. Field crops suffer from anaerobic conditions, leading to oxygen stress, CO_2 accumulation, deficiency and toxicity of certain nutrients, toxicity due to organic acids, etc. Waterlogging may cause 20–50% yield losses in cereal, pulse, and oilseed crops (wheat, barley, chickpea, lentil, mustard, etc.). A wide range of agronomic, engineering, and genetic solutions are currently available to improve the conditions of waterlogged soils and stabilize crop yields. The choice, however, depends on the cause of formation of wetlands. Some technologies are summarized below:

2.12.1.1 Removal of Excess Water

This technique works in wetlands where water accumulates due to poor drainage conditions created by one or more of the factors like low-lying fine-textured soils in high-rainfall areas, subsoil compaction caused by agricultural machinery, or natural existence of impervious hardpans (claypans, fragipans, duripans, etc.) in subsoil. Waterlogging may be a surface problem as in low-lying fine-textured soils in high-rainfall areas or subsurface problem due to shallow groundwater table. If waterlogging is a surface problem, surface drainage or drainage well methods will do. Breaking/shattering subsurface hardpans through deep tillage or using chisels, subsoilers, or rippers improves deep percolation. If waterlogging is due to shallow groundwater table, subsurface drainage is a useful drainage technique.

Bio-drainage is another way of lowering the water table in waterlogged areas. Bio-drainage is defined as "the process of removing the excess soil water through transpiration using bioenergy of the plant and radiation energy of the sun." Bio-drainage vegetation has the characteristics of fast growth rate, deep root system, and high transpiration rate. Few examples are *Eucalyptus, Populus, Casuarina, Dalbergia, Syzygium, Acacia, Prosopis, and Leucaena*. Consumptive water use of these trees varies between 6500 and 28,000 m^3/ha-year, and under ideal conditions, a tree canopy may lower water table by 1–2 m over a time period of 3–5 years (Gafni and Zohar 2001; Kapoor 2001). Bio-drainage technique has been successfully used in lowering of water table and interception of seepage from canals and channels (Sarkar et al. 2018). Bio-drainage coupled with engineering techniques gives better results.

2.12.1.2 Improving Root Zone Conditions

Such techniques work well in soils which are temporarily waterlogged during cropping season. Different options of improving root zone conditions include adoption of raised-sunken bed technology or use of ridge-furrow system of planting. Raised beds or ridges have better aeration and provide better medium for plant growth. The technique is suited to water-shy production systems. Adjusting sowing times of crops to avoid waterlogging at critical growth stages also helps in stabilizing crop yields.

2.12.1.3 Selection of Appropriate Production Systems

There are situations where drainage is rather difficult or expensive and soils remain permanently waterlogged or water saturated. Under such situations, it is better to identify water-loving production systems for cultivation, e.g., rice, sugarcane, forestry, fisheries, and duckery.

2.12.2 Water Management Under Irrigated Conditions

Rainfall is the primary source of water in agriculture. Where rainfall is insufficient to meet crop water demand, water deficit is supplemented with irrigation. The objective of water management under irrigated conditions is to maximize crop yields per unit of water use, i.e., *more crop per drop*, without wasting water resource and without compromising with soil health. The irrigation water needs to be managed most profitably at sustainable production levels. Non-judicious and unscientific use of irrigation water may lead to loss of water resource and cause land degradation through soil salinization, soil erosion, nutrient leaching, or waterlogging. The efficient irrigation system essentially has three characteristics:

(i) Minimum nonproductive water losses during conveyance of water from source to the field
(ii) Uniform water application in the field
(iii) Minimum cost of irrigation system

The traditional irrigation methods have been designed to make up the *field capacity deficit* in the root zone. The interval between two irrigations is the period taken for depletion of water in the root zone to a critical level below which root water uptake is adversely affected. Relatively large intervals between successive irrigations reduce the number of irrigations per season. It cuts down on the cost but may adversely affect the crop.

With the understanding of soil-plant-water relationships (SPAC), the irrigation management technology has received new thinking. Since biomass production is directly linked with transpiration rate, the crop must transpire at the potential rate to produce maximum dry matter. To maintain transpiration at the potential rate, soil water content must remain close to the field capacity. That would require shallow and frequent rather than deep irrigations at large intervals. Modern concept of irrigation advocates daily irrigations to maintain water in the root zone capable of supporting potential transpiration by the crop plants. It can be achieved through drip and sprinkler irrigation systems.

The basic elements of irrigation scheduling include application of water at right time, in right amount, and in right way so that the crop does not experience water stress at any time during its growth period and at the same time water is not wasted. Irrigation scheduling on scientific lines is described in detail in Chap. 3.

2.12.3 Water Management Under Rainfed Conditions

Rainfed crops depend for water entirely on rains, which are variable, uncertain, and unpredictable. Rainfed crops generally grow under water-deficit conditions, experiencing occasional water stresses of varying degrees. It results in low and unstable crop yields. Water management under rainfed situations is more difficult than under irrigated conditions. The objective of water management under rainfed situations is to *maximize agricultural (crop) productivity* per unit area. The focus of water management is to synchronize the expected water deficits with the least water-sensitive growth stages of the crop. Modeling techniques can be used to develop functional relationships between crop yields and water deficits at different crop growth stages.

The water management under water-deficit conditions is approached in two ways: (i) augmenting water resources and (ii) optimally utilizing water to achieve maximum crop productivity.

2.12.3.1 Augmenting Water Resources

Water resources can be augmented under rainfed situations by conserving/harvesting/storing rainwater *in situ* or *ex situ* and reducing the nonproductive water losses such as runoff, evaporation, and seepage. Augmentation of water resources may be achieved in different ways, grouped under four categories:

(a) Rainwater harvesting
(b) Groundwater recharging
(c) Soil moisture conservation
(d) Rejuvenation of traditional water resources

(a) *Rainwater harvesting*: The *rainwater harvesting* (or water harvesting) refers to the collection of rainwater/spring water/runoff/stream flows/rooftop water and storing it in suitable reservoirs. The rainwater may be harvested within fields (*in situ* water harvesting) or elsewhere but within the same basin/command area (*ex situ* water harvesting). The *in situ* water harvesting is achieved through land leveling, embankment of fields, and land shaping such as contour bunding and terracing. Digging trenches of suitable size, depending on the slope and amount of rainfall, across the slope or across the water flow lines or above the tree lines improves rainwater interception, profile recharging, and water availability to trees in hilly areas. Inter-plot water harvesting by proving a gentle slope to the field and harvesting runoff in a small pond at one corner of the area, or diverting runoff from one plot to another plot, or changing the configuration of the flat land into ridge-furrow system or alternate raised-sunken bed system is another effective way of *in situ* rainwater harvesting (Sharma 2003).

In *ex situ* water harvesting, the water is harvested and stored at a place away from the place of utilization but within the same basin or command area. For *ex situ* water harvesting, a suitable storage structure and water conveyance system are necessary. Water harvesting may be done on *macroscale*, i.e., watershed management on community basis, or on *microscale*, i.e., water harvesting in small reservoirs (100–200 m^3) on farmers' fields. The watersheds may vary in size between <1000 ha and >50,000 ha; agricultural watersheds are generally <1000 ha. Water harvesting and conservation at different spatial scales (i.e., at catchment, field, or microlevel) can bring sustainability to the water sector and, consequently, increase water availability in drought years.

Table 2.9 Groundwater recharging methods for rural and urban areas

Urban areas	Rural areas
Rooftop rainwater/runoff harvesting through • Recharge pit • Recharge trench • Tube wells/hand pumps • Recharge well	Rainwater harvesting through • Gully plug • Contour bund • Gabion structure • Percolation tanks • Check dam/cement plug/*Nala bund* • Recharge shaft • Dug-well recharge • Groundwater dams/subsurface dyke

(b) *Groundwater recharging*: Groundwater is an important source of water for domestic, commercial, and agricultural use. The groundwater accounts for nearly 23% of planet's freshwater resource. In many countries, the groundwater meets at least half of the total irrigation requirement of the region. For sustaining the groundwater resources, the use of groundwater should be so regulated so as not to exceed the recharging possibilities. Recharging groundwater is an important component of water management in rainfed areas. Good review papers have been published on artificial groundwater recharging (Kavuri et al. 2011; Mukherjee, 2016; Nandanwar et al. 2020). Some techniques of groundwater recharging currently available for rural and urban areas are listed in Table 2.9.

(c) *Soil moisture conservation*: The objective is to increase soil moisture storage, decrease evaporation losses, and enhance duration of water availability to crops. Several options of *in situ* soil moisture conservation are available.

- *Mulching*: Mulching is an age-old practice of soil moisture conservation. Several organic and inorganic materials of natural (field crop residues, waste plant biomass, tree leaves, farm weeds, wood chips, saw dust, coco-coir, etc.) or synthetic origin (plastic sheets, soil conditioners), soil (dust mulch), or pebbles (stone mulch) are in use as mulch. The choice of the material may depend on factors like cost, availability, ease of handling, and efficacy.
- *Alley cropping*: It is an agroforestry system in which food crops are grown in alleys formed by hedgerows of trees and shrubs. Pruning of hedgerow trees/shrubs provides excellent organic mulch material. The technique can be easily adopted by resource-poor farmers in hills and in marginal areas.
- *SOM buildup*: Organic matter (OM) plays three important functions in soil: nutritional function (supplies essential nutrients), biological function (food for microbes), and soil physical, chemical, and physicochemical function (improvement in soil properties). Water-holding capacity (WHC) and plant available water capacity (PAWC) of soils are directly linked to SOM. Studies have shown that 1% increase in SOM increases field capacity by about 2.2%, permanent wilting point by about 1%, and PAWC by about 1.5%. Soil incorporations of organic material such as FYM, compost, green manures, waste plant residues, farm weeds, or crop residues improve SOM and soil water retention. *Conservation tillage* helps in the preservation of SOM.

(d) *Rejuvenation of traditional water sources*: Traditionally, communities have been managing water resources at their own level using their traditional ecological knowledge. Several kinds of rainwater harvesting systems were developed befitting the ecological conditions across the globe. But with the implementation of government schemes for domestic water supply in many countries, some of the traditional systems were neglected. In regions where formal institutional control of water resources is limited due to a lack of coordination or stakeholder involvement, communities still rely on traditional knowledge to manage common-pool resources. However, with increasing demand for freshwater by increasing population and different economy sectors, depleting groundwater resources, recurring droughts, climate change, etc., traditional rainwater harvesting systems are attracting growing attention, especially in developing countries. It is important to protect the traditional water harvesting structures and to rejuvenate the damaged/dried/polluted water resources/structures (water springs, ponds, wells, stepwells, traditional water conveyance systems, etc.).

2.12.3.2 Optimal Utilization of Conserved Water

The moisture conserved in soil or water harvested in reservoirs is a precious commodity in water-deficit areas and hence must be utilized very carefully and efficiently. Efficient utilization of conserved water resources involves implementation of improved agronomic practices to increase crop water productivity at the field/farm level, and critical planning and implementation at basin/sub-basin level for maximizing water productivity.

Improved agronomic practices include proper choice of tillage practices, crops, and crop cultivars; adoption of mechanized farming; efficient irrigation methods; protected agriculture; mixed farming; crop rotations; integrated nutrient management; crop protection; weed control, etc. High-yielding, less water-requiring, short-duration crops having low transpiration and deep and well-branched root system are most suited to water-deficit areas. Pressurized irrigation systems (sprinkler, drip, fertigation) are the best choice in water-deficit regions. Protected agriculture has the potential of water saving by 30–60%.

2.13 Water Management in Low-Permeable Clayey Soils

Low infiltration rates and excessive runoff are serious problems in fine-textured clayey soils. Any practice that improves soil aggregation and aggregate stability increases water infiltration, decreases runoff, improves drainage, and increases soil water retention. Increased applications of organic materials to soil and conservation tillage usually improve infiltration.

2.14 Water Management in Highly Permeable Sandy Soils

The water management problems in sandy soils include high permeability, low water retention, and excessive leaching of nutrients. These challenges can be addressed in different ways.

- *Mixing fine-textured material with the root zone soil:* If the sandy soil is underlain by a fine-textured soil layer within the reach of tillage equipment, deep tillage can be used to mix the fine material with the coarse material. It may increase clay content in the root zone by 5–15% (Unger et al. 1981). The other possibility of changing the texture of surface soil layer is the mixing of fine-textured soil imported from nearby area. The technique, although expensive, has permanent/long-lasting effects. Long-term use of muddy water (water having heavy load of clay and silt) for irrigation may also gradually change soil texture towards fineness.
- *Incorporation of organic materials:* A regular addition of organic materials (OMs) in sandy soils improves soil structure and water-holding capacity of soils and lowers soil permeability.
- *Installation of subsurface barriers:* Establishing water impervious barriers below the root zone is another way of reducing percolation losses in sandy soils. Use of asphalt or plastic barriers at 30–60 cm soil depth has been found useful in checking percolation losses of water and nutrients and increasing yields of crops like vegetables (cabbage, cucumbers, tomato, potato, beans, etc.) corn, and rice. The depth of barrier application may depend on the type of crop, the equipment used, and the cost of barrier application. Sharma et al. (1995) nearly doubled grain yield of rainfed rice and water-use efficiency in a loamy sand soil through subsoil compaction achieved with a 12-t road roller (Table 2.10).

Table 2.10 Effect of subsoil barriers on rice grain yield and water-use efficiency in a loamy sand soil under rainfed conditions (Sharma et al. 1995)

	Grain yield (Mg/ha)		Water-use efficiency (kg/ha-mm)	
Treatment	1992	1993	1992	1993
Subsoil compaction[a]	4.0a	2.1a	5.06	4.34
Plastic sheet barrier[b]	2.9bc	1.9ab	3.72	3.96
Puddling	2.5c	1.1cd	3.19	2.31
Dry tillage[c]	2.4c	0.9d	3.10	1.80

Note: In a column, means followed by a common letter are not significantly different at $p = 0.05$ by DMRT

[a]Soil compaction with six passes of 12-ton road roller (Dynapac CA25) with vibration on soil surface *fb* 20 cm deep tillage

[b]Plastic sheet placed at 40 cm depth after removing topsoil

[c]Soil was disc-plowed dry to 15 cm depth

2.15 Question Bank

2.15.1 *Short Questions:*

i. How is water significant for plants ecologically and physiologically?
ii. What important physiological changes occur in plants due to waterlogging?
iii. In what way waterlogging affects soil fertility?
iv. Briefly explain morphological and physiological effects of water stress on plant growth.
v. Explain briefly as to how the degree of water stress affects plant growth.
vi. Define SPAC. How is this concept important in soil-plant-water relations?
vii. What is the impact of soil water regime on the nutrients' transformation and nutrients' translocation in soil?
viii. Distinguish between "crop water use" and "crop water requirement."
ix. Distinguish between "water-use efficiency" and "water productivity."
x. Enlist different methods of determining evapotranspiration (ET).
xi. What is the role of "crop coefficient" in estimating crop ET?
xii. Is it possible to increase crop WUE through irrigation management? If yes, how? If no, why?
xiii. What is rainwater harvesting? What is its role in augmenting water resources in a region?
xiv. Briefly explain how rejuvenation of traditional water sources is important in rainfed/dryland agriculture.
xv. Is it possible to decrease the permeability of sandy soils? If yes, how? If not, why?
xvi. In what way remote sensing is better than conventional methods of measurements?

2.15.2 *Briefly Explain Why?*

i. The distribution of vegetation over the surface of the earth is largely controlled by the availability of water.
ii. Water is essential for maintaining turgidity in plants.
iii. Waterlogging may reduce crop yields by 20–50% or even more.
iv. Water stress induces early senescence in plants.
v. Water-stressed plants are usually stunted in growth.
vi. SPAC is an important concept in soil-plant-water relations.
vii. Water and nutrients have synergistic effect on WUE and NUE.
viii. Nutrients' translocation in soil is important in nutrients' availability to plants.
ix. Most of the water needs of plants are met from top 45–60 cm soil layer.
x. The crop water use practically equals evapotranspiration; water used in metabolic processes in plant system is ignored.

xi. The climatic factors strongly influence crop ET.
xii. “Water productivity” is a better concept than “water-use efficiency” on a basin/sub-basin system basis.
xiii. Short-duration, deep-rooted, and drought-resistant crops give higher WUE.
xiv. Reduction in evaporation is a better option than reduction in transpiration to increase WUE.
xv. Deficit irrigation scheduling helps in improving WUE.
xvi. Drainage improves crop productivity in waterlogged soils.
xvii. Recharging groundwater is an important component of water management in rainfed areas.
xviii. Conservation tillage improves water-holding capacity of soils.
xix. Fertigation improves WUE as well as NUE.
xx. “Rainwater harvesting” and “water harvesting” are interchangeably used terms.

2.15.3 Fill in the Blanks:

i. Water content in mature woody plant tissues ranges from …… to …….
ii. Water content in herbaceous plants ranges from …… to ……
iii. Soil saturation for 24 h may reduce the growth and yield of most of the upland crops by ……..
iv. The oxygen level in a saturated soil declines to critical levels for plant growth within ……… hours.
v. Nicotine content of tobacco leaves ……………… with an increase in water stress.
vi. Protein content of wheat grains ……………… with an increase in water stress.
vii. Cell growth and cell wall synthesis ………………… due to water stress.
viii. CO_2 metabolism is ……….. in plant tissues due to water stress.
ix. “Moderate water stress” refers to a condition when the plant water potential is by ………
x. The water in the SPAC flows in response to ………………
xi. The total potential difference between soil and atmosphere in arid climates may cross ……..
xii. Root interception contributes only about ……….. of total nutrient absorption by plant roots.
xiii. …………… supplies most of the N, S, Ca, and Mg, and to some extent K, to plant roots.
xiv. Most of the P and K reach plant roots through …………
xv. In a uniform soil profile with adequate moisture, crop plants usually extract …….. moisture from the first quarter of the root zone.
xvi. Only about ……… of the total water absorbed by the plant is used in metabolic processes of the plant.

xvii. One hectare of corn crop may transpire between …………………. liters of water per day.

xviii. …………………. refers to "the quantity of water transpired by a crop to produce a unit amount of dry matter."

xix. ……………….. is the reciprocal of transpiration ratio.

xx. The ratio of leaf surface area to land surface area is known as ……….

xxi. The consumptive water use refers to the amount of water used in the metabolic process of the plant and …………

xxii. Observation and measurement of objects/parameters from a distance is called …………

xxiii. …………… is defined as "the ratio of the amount of water consumed by the crop (ET) to the amount of water supplied through irrigation."

[**Key:** (i) 45 to 50% (ii) 70 to 90% (iii) 50% (iv) 48–96 (v) increases (vi) decreases (vii) decrease (viii) inhibited (ix) –12 to –15 bar (x) energy differences/potential gradient (xi) 1000 bars (xii) 3% (xiii) Mass flow (xiv) diffusion (xv) 40% (xvi) 0.5–1.0% (xvii) 28,000 and 37,000 (xviii) Transpiration ratio (xix) Water use efficiency (xx) leaf area index (xxi) transpiration (xxii) remote sensing (xxiii) Irrigation efficiency]

2.15.4 State Whether the Following Statements Are True (T) or False (F):

i. Hydraulic conductivity of plant roots is high in waterlogged soils.

ii. Oxygen depletes faster in waterlogged OM-rich soils.

iii. Anoxia is responsible for the production of phytotoxins in soil.

iv. Plants usually suffer from K deficiency in waterlogged soils.

v. Waterlogging enhances mineralization in soils leading to N deficiency.

vi. Waterlogging decreases the sensitivity of plant roots to high temperatures.

vii. Waterlogging decreases root hydraulic conductivity.

viii. Excess water tolerance of crops increases in the order: oats>wheat>barley.

ix. Water stress induces early senescence in plants due to accumulation of ascorbic acid.

x. Water stress decreases wheat yield but increases protein content of wheat grains.

xi. Nicotine content of tobacco leaves increases with an increase in water stress.

xii. Water stress inhibits CO_2 metabolism in plant tissues.

xiii. Water in SPAC flows in response to differences in water content.

xiv. The total potential difference between soil and atmosphere may amount to hundreds of bars and may cross 1000 bars in arid climates.

xv. Plant roots grow to access around 30% of the available nutrients in soil.

xvi. Most of the P and K reach plant roots through diffusion.

xvii. Most of the N, S, Ca, and Mg reach plant roots through mass flow.

xviii. Most of the water needs of plants are met from the plough layer.

xix. Nearly 70% of water requirement of plants is met from the top two quarters of the root zone.
xx. "Crop water requirement" and "total crop water requirement" are many times considered as synonymous terms.
xxi. Consumptive water use refers to the fraction of total water absorbed by the plant, which is used in metabolic processes.
xxii. "Water-use efficiency" and "water productivity" are synonymous terms.
xxiii. "Water-use efficiency" is the reciprocal of "transpiration ratio."
xxiv. The ratio of leaf surface area to land surface area is known as leaf area index.
xxv. The actual crop ET can be measured using a lysimeter.
xxvi. Crop coefficient is used to compute actual crop ET from the reference ET value.
xxvii. Crop coefficient varies with the crop species but is independent of crop stage.
xxviii. The value of crop coefficient increases with the crop stage.
xxix. Deficit irrigation scheduling is a way of increasing crop WUE.

[**Key:** (i) F (ii) T (iii) T (iv) F (v) F (vi) F (vii) T (viii) T (ix) F (x) F (xi) T (xii) T (xiii) F (xiv) T (xv) F (xvi) T (xvii) T (xviii) F (xix) T (xx) T (xxi) F (xxii) F (xxiii) T (xxiv) T (xxv) T (xxvi) T (xxvii) F (xxviii) F (xxix) T]

2.15.5 Multiple-Choice Questions:

i. The plant tissue mass has water content around

   a. 20–35%
   b. 40–55%
   c. 60–85%
   d. 80–95%

ii. Plant growth is adversely affected due to waterlogging as

   a. Water is toxic to plants
   b. Water produces anoxia in soil
   c. Water destroys soil structure
   d. Water leaches down plant nutrients

iii. Soil saturation for the following duration may reduce the growth and yield of most of the upland crops by 50%:

   a. 8 h
   b. 16 h
   c. 24 h
   d. 48 h

iv. Which one of the following is not correct:

   a. Waterlogging causes oxygen deficiency.

b. Waterlogging decreases soil redox potential.
c. Waterlogging increases iron toxicity.
d. Waterlogging increases root hydraulic conductivity.

v. Water stress induces early senescence in plants due to

a. Accumulation of abscisic acid
b. Accumulation of ascorbic acid
c. Accumulation of folic acid
d. Accumulation of malic acid

vi. Only a small fraction of absorbed water is used in metabolic processes of the plant, which is equal to

a. 0.1–0.5%
b. 0.5–1.0%
c. 1.0–5.0%
d. 1.0–10.0%

vii. Consumptive water refers to

a. Metabolic water + irrigation
b. Metabolic water + evaporation
c. Metabolic water + evapotranspiration
d. Metabolic water + transpiration

viii. Root interception is primarily responsible for the uptake of

a. N
b. P
c. K
d. Mg

ix. Most of the P and K reach plant roots through

a. Root interception
b. Mass flow
c. Diffusion
d. Root interception and mass flow

x. The contribution of diffusion in the uptake of P by plant roots is around

a. 35%
b. 55%
c. 75%
d. 95%

xi. In a uniform soil profile with adequate moisture, the water extraction by plant roots from the first quarter of the root zone is around

a. 20%
b. 40%

c. 60%
d. 80%

xii. The value of crop coefficient generally varies between

a. 0.2 and 0.5
b. 0.2 and 0.7
c. 0.2 and 0.9
d. 0.2 and 1.2

[**Key:** (i) d (ii) b (iii) c (iv) d (v) a (vi) b (vii) d (viii) d (ix) c (x) d (xi) b (xii) d]

References

Blaney HF, Criddle WD (1950) Determining water requirements in irrigated areas from climatological and irrigation data. US Soil Conserv Ser Tech Publ 96

Briggs LJ, Shantz HL (1913) The water requirement of plants. In: Bureau of plant industry bulletin. USDA, Washington, DC, pp 282–285

Christiansen JE (1968) Pan evaporation from climate data. Am Soc Civil Eng Proc J Irrig Drain Div 1968(194):243–265

Colmer TD, Greenway H (2011) Ion transport in seminal and adventitious roots of cereals during O_2 deficiency. J Exp Bot 62:39–57

Dardanelli JL, Bachmeier OA, Sereno R, Gil R (1997) Rooting depth and soil water extraction patterns of different crops in a silty loam Haplustoll. Field Cropos Res 54(1):29–38

Evans RG (1999) Irrigation choices for growing *Vitis vinifera* grapes in central Washington. In: Watson J (ed) Proceedings of growing grapes in Eastern Washington symposium. Good Fruit Grower, Yakima, WA

Evans RG, Sadler EJ (2008) Methods and technologies to improve efficiency of water use. Water Resources Res 44(7):15

FAO. Food and Agriculture Organization of the United Nations 2015. Available at: http://www.fao.org/3/a-bc600e.pdf

Gafni A, Zohar Y (2001) Sodicity, conventional drainage and bio-drainage in Israel. Aust J Soil Res 39:1269–1278

Gregory PJ, Simmonds LP, Warren GP (1997) Interactions between plant nutrients, water and carbon dioxide as factors limiting crop yields. Philosophical Trans Royal Soc B: Bio Sci 352:987–996

Hati K, Mandal K, Misra A, Ghosh P, Bandyopadhyay K (2006) Effect of inorganic fertilizer and farmyard manure on soil physical properties, root distribution, and water-use efficiency of soybean in Vertisols of central India. Bioresource Tech 97:2182–2188

Havlin JL, Beaton JD, Tisdale SL, Nelson WL. Soil fertility and fertilizers: an introduction to nutrient management, 6th ed. 2005; Prentice Hall, Upper Saddle River, NJ.

Hsiao TC (1973) Plant responses to water stress. Ann Rev Plant Physiol 24:519–570

Hu R, Wang X, Pan Y, Zhang Y, Zhang H (2014) The response mechanisms of soil N mineralization under biological soil crusts to temperature and moisture in temperate desert regions. European J Soil Bio 62:66–73

IPCC (2014) Climate change 2014: synthesis report. In: Contribution of working groups I, II and III to the Fifth Assessment Report of the Intergovernmental Panel on Climate Change. IPCC, Geneva, p 10

Israelsen OW (1932) Irrigation principles and practices. Wiley and Sons, New York, p 411

Jensen ME (1977). Water conservation and irrigation systems: climate. In: Technical seminar proceedings, Columbia, Missouri, pp. 208–250.

Kapoor AS (2001) Biodrainage: a biological option for controlling waterlogging and salinity. Tata McGraw – Hill Publishing Company Limited, New Delhi, p 315

Kapur OC, Aggarwal GC, Kanwar BS, Tripathi PR (1978) Effect of mulching with pine needles on the yield of timely and late sown wheat. Indian J Agric Sci 48:605–609

Kavuri M, Boddu M, Annamdas VGM (2011) New methods of artificial recharge of aquifers: a review. In: Proceedings of 4th International Perspective on Water Resources & the Environment 2011 (January 4-6, 2011). National University of Singapore (NUS), Singapore

Kijne JW, Barker R, Molden D. (eds.). Water productivity in agriculture: limits and opportunities for improvements. In: Comprehensive assessment of water management in agriculture, series 1, 2003a, CABI International, UK.

Kijne JW, Tuong TP, Bennett J, Bouman BAM, Oweis T (2003b) Ensuring food security via crop water productivity improvement. In: Background papers: challenge program for food and water. CGIAR-IWMI, Colombo, Sri Lanka, pp 1–42

Lawes JB (1850) Experimental investigation into the amount of water given. J Hortic Soc London 5:38–63

Manik SMN, Pengilley G, Dean G, Field B, Shabala S, Zhou M (2019) Soil and crop management practices to minimize the impact of waterlogging on crop productivity. Front Plant Sci 10:140

Martin J, Leonard W, Stamp D (1976) Principles of field crop production, 3rd edn. Macmillan Publishing Co, New York

Mehta R, Pandey V (2016) Crop water requirement (ETc) of different crops of middle Gujarat. J Agromet 18(1):83–88

Molden DJ, El Kady M, Zhu Z (1998) Use and productivity of Egypt's Nile. In: Paper presented at the 14th Technical Conference on Irrigation, Drainage and Flood Control, 3–6 June 1998. USCID, Phoenix, AZ

Molden DJ, Oweis T, Steduto P, Kijne J (2007) Pathways for increasing agricultural water productivity. In: Water for food water for life, vol 2007. Earthscan Publ, UK, pp 279–310

Mukherjee D (2016) A review on artificial groundwater recharge in India. Int J Civil Eng 3(1):57–62

Munns R, Sharp RE (1993) Involvement of abscisic acid in controlling plant growth in soil of low water potential. Funct Plant Biol 20(5):425–437

Nandanwar AA, Raut AY, Qureshi AS, Bande CM, Dhawase JM, Khobragade MM (2020) A review paper on groundwater recharge by utilizing waste water from residential area. Int Res J Eng Tech 7(2):368–370

Penman HL (1948) Natural evaporation from open water, bare soil and grass. Proc R Soc London Ser A 193:120–146

Philip JR (1966) Plant water relations: some physical aspects. Annu Rev Plant Physiol **17**:245–268

Rajanna GA, Dass A, Paramesha V (2018) Excess water stress: effects on crop and soil, and mitigation strategies. Popular Kheti 6(3):48–53

Ramsar Convention on Wetlands (2018) Global wetland outlook: state of the world's wetlands and their services to people. Ramsar Convention Secretariat, Gland, Switzerland

Sarkar A, Banik M, Ray R, Patra S (2018) Soil moisture and groundwater dynamics under biodrainage vegetation in a waterlogged land. Int J Pure Appl Biosci 6:1225–1233

Seckler D, Molden D, Sakthivadivel R (2003) The concept of efficiency in water-resources management and policy. In: Kijne JW, Barker R, Molden D (eds) Water productivity in agriculture: limits and opportunities for improvement. CABI Publishing, CAB International, USA, pp 37–51

Sharma PK (1989) Effect of periodic moisture stress on water-use efficiency in wetland rice. Oryza 26:252–257

Sharma PK (2003) Raised-sunken bed system for increasing productivity of rice-based cropping system in high rainfall areas of Himachal Pradesh. J Indian Soc Soil Sci 51:10–16

Sharma PK, Bhushan L (1999) Effect of soil-water regime on evapo-transpiration and dry matter production of two rice cultivars. J Indian Soc Soil Sci 47:353–355

Sharma PK, Kharwara PC (1990) Soil-stored available water and seasonal rainfall as an index of success or failure of rain-fed crops. Indian J Agric Sci 60:165–168

Sharma PK, Ingram KT, Harnpichitvitaya D (1995) Subsoil compaction to improve water use efficiency and yield of rainfed lowland rice in coarse-textured soils. Soil Till Res 36:33–44

Sinclair TR, Tanner CB, Bennett JM (1984) Water-use efficiency in crop production. Bio Sci 34(1):36–40

Singh B, Ryan J, Campbell C, Kröbel R (2015) Nutrient management and water use efficiency for sustainable production of rain-fed crops in the World's dry areas. In: Heffer DP, Magen H, Mikkelsen R, Wichelns D (eds) Managing water and fertilizer for sustainable agricultural intensification. International Fertilizer Industry Association (IFA), International Water Management Institute (IWMI), International Plant Nutrition Institute (IPNI), and International Potash Institute (IPI), Paris, France, pp 140–167

Stanhill G (1986) Water use efficiency. Adv Agron 39:53–85

Steduto P (1966) Water use efficiency. In: Pereira LS et al (eds) Sustainability of irrigated agriculture, NATO ASI Ser E, vol 312. Kluwer Acad, Dordrecht, Netherlands, pp 193–209

Sun Z, Zheng J, Sun W (2009) Coupled effects of soil water and nutrients on growth and yields of maize plants in a semi-arid region. Pedosphere 19:673–680

Tandon HLS (1987) Phosphorus in dryland agriculture. In: Tandon HLS (ed) Phosphorus research and agricultural production in India. Fertilizer Development and Consultation Organisation, New Delhi, pp 73–84

Thornthwaite CW (1948) An approach toward a rational classification of climate. Geogr Rev 38:55–94

Unger PW, Eck HV, Musick JT (1981) Alleviating plant water stress. In: Arkin GF, Taylor HM (eds) Modifying the root environment to reduce crop stress, vol 1981. St. Joseph, Mich. American Society of Agricultural Engineers, pp 61–96

Verhoeven JTA, Setter TL (2010) Agricultural use of wetlands: opportunities and limitations. Ann Botany 105:155–163

Yadav BL, Yadav CP (2008) Effect of Nitrogen Fertilization on Soil-Plant-Water Relationship under Different Soil Water Conservation Practices in Mustard. J Indian Soc Soil Sci 56:114–117

Yadav RL, Singh SR, Prasad K, Dwivedi BS, Batta RK, Singh AK, Patil NG, Chaudhari SK (2000) Management of irrigated agro-ecosystem. In: Natural Resource Management for Agricultural Production in India (JSP Yadav and GB Singh, Eds.), Indian Soc Soil Sci, New Delhi:775–870

Chapter 3
Irrigation Management

Major part of the crop water requirement is met through precipitation, i.e., rainfall and snowfall. To meet the remaining water requirement, water is applied artificially to crops, a process called *irrigation*. Irrigation management refers to regulation of irrigation water in such a way so as to maximize irrigation efficiency without wasting water, nutrients, soil, and energy and without sacrificing crop yields. Different crops have different water requirements. Irrigation is applied to crops as per their water needs, and in amounts and rates consistent with soil's infiltrability and water-holding capacity.

In this chapter, we shall discuss irrigation and different components of irrigation management, viz. irrigation scheduling, irrigation efficiency, fertigation, irrigation management in saline soils and highly permeable soils, and use of saline waters for irrigation.

3.1 Irrigation

Although irrigation has been defined differently by different people and organizations, the basic concept of irrigation, however, remains the same. *Irrigation* may be defined as "the process of supplying water to land by artificial means to supplement the natural supply of water." "The regulation and monitoring of water applications to plants" are known as *irrigation water management*.

The *objective(s)* of irrigation in agriculture may be one or more of the following:

- To meet the water deficit in the root zone essential for normal plant growth and development
- To facilitate nutrient application (*fertigation*) and enhance nutrient availability to plants
- To maintain soil moisture for optimizing microbial activities and soil fertility

P. K. Sharma, S. Kumar, *Soil Physical Environment and Plant Growth*,
https://doi.org/10.1007/978-3-031-28057-3_3

- To dilute or leach excess salts, toxins, or pollutants from the root zone
- To soften the land for facilitating tillage or intercultural operations
- To reduce soil penetration resistance for easy root growth
- To protect plants against frost injury

The concept of irrigation dates back at least 8000 years, but the earliest known systems of irrigation began in 6000 BC in Egypt and Mesopotamia. The canals dug by the Sumerians are considered as the first ever works of engineering. A complex system of underground canals and artificial reservoirs for storing water was built around 300 BC in Sri Lanka, which is still in existence today. Sri Lankans were the first to build true water reservoirs to store water extracted from the ground for use as irrigation. The *qanats*, developed in ancient Persia (present-day Iran) about 800 BC, are among the oldest known irrigation methods in use today. Irrigation in Mexico dates back to about 600 BC.

Surface flooding is an ancient technique of water application in plain areas world over. Terrace irrigation is the oldest method of irrigation in hilly regions all over the world, including in Peru, Syria, China, and India. The remains of terrace irrigation system developed around 4000 BC can still be seen in the Zaña Valley of Peru. Modern irrigation systems started taking shape in 1800s.

Although irrigation has been practiced worldwide for more than 6000 years, more innovation has occurred in this sector during the last about 100 years (Postel, 1999). Almost every aspect of irrigation has seen significant innovation, including diversion works, pumping, filtration, conveyance, distribution, application methods, drainage, power sources, scheduling, fertigation, chemigation, erosion control, land grading, soil water measurement, and water conservation. Such innovations have led to about fivefold increase in irrigated area globally. Area equipped for irrigation (AEI) increased from about 63 million ha in 1900 to more than 306 million ha in 2005 (Siebert et al. 2015).

3.2 Concept of Irrigation Water Management

The focus of an efficient irrigation management system is to supply water to the root zone in optimum amount, avoiding over- or under-irrigation. The proper irrigation water management results in:

- Preventing under- or overuse of irrigation water
- Preventing irrigation-induced erosion
- Reducing operational cost of irrigation including labor, pumping, energy, etc.
- Maintaining or improving quality of groundwater and downstream surface water
- Increasing crop biomass production and product quality

The proper irrigation management requires knowledge of crop water needs, water source, and net irrigable area with the given water source. The following equation is the basic tool of irrigation water management:

$$QT = AD_i \tag{3.1}$$

where Q is the flow rate (m^3/min), T is the time (h), A is the area to be irrigated (m^2), and D_i is the depth of irrigation water (m) to be applied. The following example will make the concept clear.

Numerical example: Calculate the time required to apply 6 cm of water to an area of 1.0 ha if the flow rate of water is 3.0 m^3/min.

Solution:

Given: Water flow rate (Q) = 3.0 m^3/min

Area to be irrigated (A) = 1.0 ha (i.e., 10,000 m^2)

Depth of irrigation (D_i) = 6.0 cm (i.e., 0.06 m)

Time of water application (T) = ?

Using Eq. 3.1

$$QT = AD_i$$

$$\text{or}\, T = AD_i / Q$$

$$\text{i.e.}\, T = (10{,}000 \times 0.06) / 3.0 = 200\,\text{min} = 3\,\text{h}\,20\,\text{min}$$

3.3 Irrigation Scheduling

Irrigation scheduling refers to "the process of determining when, how, and how much water to be applied to the field." The aim of irrigation scheduling is to apply water at the right time, in the right amount, at the right place, and in a right way to achieve optimum economic crop yields. Enough water must be applied to fully wet the plant's root zone while minimizing overwatering. The overwatering causes wastage of water resource, decreases water-use efficiency, increases chances of nutrient leaching and pollution of groundwater, and may create oxygen stress in the root zone and increase cost of irrigation. The under-watering exposes the crop to water stress and may harm the crop growth and result in loss of economic yield. Under both the situations, water-use efficiency and nutrient-use efficiency are low. Some crop growth stages are very sensitive to water stress, and even a mild water stress may cause significant reduction in crop yield.

The *objectives* of irrigation scheduling (i.e., irrigation water management) include:

- Optimization of water resources
- Soil water regulation to obtain desired crop response
- Preventing economic yield losses of crops due to water stress
- Maximizing use efficiency of agricultural inputs
- Minimizing irrigation-induced soil erosion hazards

- Minimizing leaching of fertilizer salts and other agrochemicals beyond the root zone
- Optimizing air-water balance in the root zone

In order to determine when and how much to irrigate, information on plant available water capacity (PAWC), permissible water depletion limit (P) or plant water status, rooting depth, plant growth stages critical for water stress, etc. is required. The thumb rule for most of the crops is to irrigate when half of the PAWC in the root zone is depleted. Equally important is to select an irrigation method to achieve the highest irrigation efficiency and crop water-use efficiency. Different methods used to determine when, how much, and how to irrigate are discussed in the following sections.

3.4 Basic Elements of Irrigation Scheduling

There are three basic elements of irrigation scheduling:

(i) Timing of irrigation, i.e., *when to irrigate*?
(ii) Amount of irrigation, i.e., *how much to irrigate*?
(iii) Method of irrigation, i.e., *how to irrigate*?

For optimum irrigation water management, the irrigation manager must have answers to these three questions. These answers depend on several soil (physical properties, viz. topography, soil depth, soil texture, infiltration, pore size distribution, bulk density, water-holding capacity, and soil water regime, and chemical properties, viz. soil salinity and alkalinity/sodicity), crop (crop type, crop density, vegetative cover or leaf area index, and crop growth stage), irrigation water quality (electrical conductivity EC, sodium adsorption ratio SAR, residual sodium carbonate RSC, and ion toxicity—sodium, boron, fluoride, heavy metals, etc.) and quantity, and climate parameters (rainfall, temperature, relative humidity, wind speed, day length, and solar radiation). The cost of irrigation (labor, pumping, etc.) also plays an important role in deciding irrigation methodology.

3.4.1 Timing of Irrigation

To accumulate maximum biomass, the crop must transpire at the potential rate. The optimum time of irrigation is one when the crop ET starts falling below its potential rate (ET_p). Different approaches have been investigated to determine the optimum time of irrigation using indices based on soil, crop, and/or environmental conditions. Each one of them has its own advantages and disadvantages. In this section, we shall briefly discuss these approaches. For details, one may refer to basic books on irrigation.

3.4.1.1 Irrigation Scheduling Based on Plant Characteristics

- *Crop appearance*: It is a qualitative approach in which time of irrigation is decided based on the appearance of the crop, viz. color of foliage, drooping, wilting, and rolling of leaves, stunted growth, etc. It is a practical but not very scientific approach. The observations may vary with the observer and experience. There may be a personal error in judgment. Sometimes, observations are misleading or confusing with nutritional disorders or insect attacks, etc. In some crops, the wilting signs are not very conspicuous. Further, by the time the symptoms appear, the damage to the crop due to moisture stress may have already occurred.
- *Physiological stages of crop*: Some physiological growth stages of crops are very sensitive to moisture stress and are called *critical growth stages* for irrigation. The crops are irrigated at these physiological stages. The critical growth stages vary with the crop species. The critical growth stages of some crops are shown in Table 3.1. It may be noted that irrigation with prefixed amount of water at these stages may result in over-irrigation if the occasional rainfall events between two irrigations are ignored.
- *Plant water status*: Plant water status may be determined as *relative leaf water content* or *leaf water potential or leaf diffusion resistance*. The threshold water

Table 3.1 Growth stages of some crops critical for irrigation

Crop	Critical growth stage for irrigation
Rice	Panicle initiation, flowering
Wheat	Crown root initiation, shooting, earhead formation
Sorghum	Booting, flowering, milky and dough stage
Maize	Tasseling, silking, early grain formation
Pearl millet	Heading, flowering
Finger millet	Panicle initiation, flowering
Groundnut	Flowering, peg penetration, seed development
Sunflower	Two weeks before flowering to 2 weeks after flowering
Cotton	Flowering, boll development
Chilies	Flowering
Sugarcane	Formative stage
Pulses	Flowering, pod formation
Soybean	Blooming, seed formation
Tobacco	Immediately after transplanting, knee stage
Citrus	Fruit setting and enlargement stage
Banana	Early vegetative period, flowering, fruit formation/ development
Tomato	From the commencement of fruit set
Potato	Tuber initiation to tuber maturity
Cabbage	Head formation until becomes firm
Carrot	Root enlargement

content values in plant tissues below which important physiological and growth phenomena are adversely affected serve as reliable indices for scheduling irrigation. Determination of these indices involves technical know-how, which may be difficult for the farmers. This technique is good for researchers but beyond the scope of farmers.

- *Plant canopy temperature*: The canopy temperature of a crop is closely related to the crop ET. The ET is an endothermic process. It keeps canopy temperature low as compared to the ambient temperature. Any reduction in ET leads to rise in canopy temperature. Consequently, canopy temperature of a well-watered crop is lower than the water-stressed crop. The difference in canopy temperature may be as high as 8 °C or even more. This concept has been used for deciding the time of irrigation.

The canopy temperature is measured by using specially designed infrared thermometers. The temperature difference (ΔT) between stressed and unstressed crop or between crop canopy and surrounding air temperature has been used as an index for scheduling irrigation. Canopy temperature > air temperature indicates crop water stress and calls for irrigation, while canopy temperature < air temperature indicates sufficient crop water content. The *stress degree-days* (SDD) index, which refers to the summation of ΔT over a period, has also been used for determining the timings and amounts of irrigation to crops.

3.4.1.2 Irrigation Scheduling Based on Soil Parameters

- *Soil water content*: Around 20–70% fraction of plant available soil water content (PAWC) is easily absorbed by the plants, without creating plant water stress. This fraction of PAWC is called *readily available water*. Below this *threshold* water content, the water availability to plants starts declining progressively and the plants have to spend more and more energy to absorb water. Ultimately, a stage is reached when plant roots fail to absorb water. Irrigation must be applied before *threshold* water content is reached. The threshold water content is crop specific.

The limit up to which soil water may be allowed to deplete safely without any adverse effect on plant growth is known as *maximum allowable depletion (MAD)/ permissible water depletion limit*. Irrigation is applied as soon as MAD limit is reached. Keeping soil water content slightly above the MAD value supports a good crop.

Different crops require irrigation at different *MAD* values, such as the following:

(a) Shallow-rooted, high-value fruit or vegetable crops need irrigation at 25–40% *MAD*.
(b) Vineyards, berries, medium-rooted row crops need irrigation at 40–50% *MAD*.
(c) Grain crops, deep-rooted row crops, forage crops, and mature trees need irrigation around 50–70% *MAD*.

The FAO-recommended MAD values are:

- 0.20–0.30 for shallow-rooted seasonal crops (e.g., vegetable crops)
- 0.40–0.60 for deep-rooted field crops (e.g., cereal crops) and mature trees

The readily available soil water content is the product of PAWC and the MAD limit. For example, if PAWC is 160 mm/m soil depth and MAD limit is 0.5, then the readily available water content would be (160 × 0.5=) 80 mm/m soil depth. The irrigation is needed as soon as 80 mm/m water is consumed by the crop or depleted from the root zone.

Soil moisture determinations should be made at 7–10-day interval during winters but more frequently during summers, and at critical growth periods of the crop. Measurements should be made in the active root zone from where plant roots extract moisture.

The tensiometer readings may also be related to MAD values using soil-moisture characteristic curve for scheduling irrigation. Some data equating matric potential with MAD for different textured soils are shown in Table 3.2.

Soil water content as an index of irrigation scheduling has certain limitations, such as the following:

- All calculations of *PAWC*, water depletion, etc. are made for a prefixed root zone depth, which may not be correct under field conditions, because the root system of crops is dynamic in nature.
- Frequent observations with respect to soil water depletion and root growth are beyond the knowledge and means of farmers; moreover, frequent soil water determinations are laborious, time consuming, and costly and may also be destructive (thermogravimetric method).
- Even a small fraction of the root system located in a wet soil layer at deeper depths may absorb enough water to meet the ET demand of the crop, although the assumed root zone may be dry enough suggesting the need for irrigation.

Table 3.2 Guidelines for using matric potential data to schedule irrigation events based on the MAD values in texturally different soils

Soil texture	Matric potential (bar) equivalent to 25% MAD	Matric potential (bar) equivalent to 50% MAD
Sand	−0.12	−0.20
Loamy sand	−0.15	−0.25
Sandy loam	−0.20	−0.40
Loam	−0.30	−0.60
Silt loam	−0.40	−0.85
Silt	−0.40	−0.85
Sandy clay loam	−0.40	−0.85
Clay loam	−0.45	−0.90
Silty clay loam	−0.45	−0.90
Sandy clay	−0.50	−0.95
Silty clay	−0.50	−0.95
Clay	−0.50	−0.95

- *Matric potential of soil water*: The matric potential in the active root zone of the crop is an important index of scheduling irrigation. Water is available to plants in the potential range of −0.1 and −15 bar, although water may not be equally available throughout this potential range. The ease of water absorption by the plant roots declines progressively as the water potential decreases. The safe limit of matric potential for applying irrigation is much below −15 bar and depends largely on the type of crop and climatic conditions, and also on the depth of measurement of matric potential. Critical matric potential values for irrigating the crops on different soils under different evaporative conditions are shown in Table 3.3. Matric potential can be easily determined by using tensiometers or electrical resistance blocks (e.g., gypsum blocks).

Table 3.3 Matric potential values critical for irrigation to some crops (Taylor and Ashcroft, 1972)

Crops		Matric potential (bars)
Vegetable crops	Beans	−0.75 to −0.20
	Cabbage	−0.60 to −0.70
	Onions	
	Early growth Bulbing time	−0.45 to −0.55 −0.55 to −0.65
	Potato	−0.30 to −0.50
	Tomato	−0.80 to −1.50
	Carrot	−0.55 to −0.65
	Broccoli	
	Early After budding	−0.45 to −0.55 −0.55 to −0.70
	Cauliflower	−0.60 to −0.70
	Sugar beet	−0.40 to −0.60
	Lettuce	−0.40 to −0.60
Grain crops	Sweet corn	−0.50 to −1.00
	Corn	
	Vegetative period Ripening period	−0.50 −8.00 to −12.00
	Small grains	
	Vegetative period Ripening period	−0.40 to −0.50 −8.00 to −12.00
Fruit crops	Oranges	−0.20 to −1.00
	Lemons	−0.40
	Grapes	
	Early season During maturity	−0.40 to −0.50 >−1.00
	Strawberries	−0.20 to −0.30
	Bananas	−0.30 to −1.50
	Cantaloupe	−0.35 to −0.40
	Deciduous fruits	−0.50 to −0.80

Note: 1. Matric potential values are for tensiometers installed at 45 cm depth
2. Where two values of matric potential are given, the higher value is for high evaporative demand and the lower value for low evaporative demand condition.

Table 3.4 Recommended tensiometer depths based on rooting depth

Rooting depth (cm)	Shallow tensiometer depth (cm)	Deep tensiometer depth (cm)
45	20	30
60	30	45
90	30	60
≥120	45	90

In this approach, two or three tensiometers are installed in the field. Two tensiometers are installed one each at one-third and two-thirds of the rooting depth. The recommended depths for different rooting zones are shown in Table 3.4. Irrigation is started when the shallow (top) tensiometer reads the desired matric potential, and irrigation is stopped when the matric potential in the second tensiometer starts rising.

Three tensiometers are installed one each in the active root zone, at the bottom of the active root zone, and midway of the two tensiometers. The water application begins when the first tensiometer reads the desired matric potential, and irrigation is discontinued when the middle tensiometer starts showing increase in matric potential. The matric potential in the tensiometer installed at the bottom of the root zone takes a day or so before it starts rising. If this tensiometer does not indicate any rise in the matric potential after irrigation, the depth of the second tensiometer may be slightly increased. The maximum matric potential in this tensiometer should not rise above the field capacity; otherwise, deep percolation of water/leaching is indicated. This method is particularly used in automatic irrigation systems.

3.4.1.3 Irrigation Scheduling Based on Climatic Parameters

- *IW/CPE ratio*: It refers to "the ratio between a fixed amount of irrigation water depth (IW) and the cumulative pan evaporation (CPE) minus *effective rainfall*[1] during the assessment period." It is a climatic approach based on the principle that for accumulating maximum biomass, plants must transpire at a potential rate. The evapotranspiration primarily depends on the type of crop, climatic conditions, and soil water regime.

In this approach, irrigation of a prefixed depth (IW) is applied when a prefixed IW/CPE ratio is reached. The ratio is fixed based on the type of crop and climatic conditions. Generally, 6–8 cm irrigation depth is applied at IW/CPE ratio varying between 0.7 and 1.0. The ratio may be high (0.9–1.0) for shallow-rooted and more water-requiring crops, such as vegetables, and low (0.7–0.8) for medium- to deep-rooted and stress-tolerant field crops, such as cereals and grasses. It is a simple

[1] The amount of rainwater retained in the root zone is called effective rainfall (R_e). If rainfall (R) >75 mm/month, $R_e = 0.8\,R$; if $R < 75$ mm/month, $R_e = 0.6\,R$.

approach and can be easily adopted under field conditions. The following example will make the concept clear.

Numerical example: What would be the frequency/timing of irrigation if 6 cm irrigation has to be applied at IW/CPE ratio of 0.8?

Solution: The timing of irrigation may be computed as follows:

$$IW / CPE = 0.8$$

$$\text{or}\, CPE = IW / 0.8$$

$$\text{or}\, \text{CPE} = 6.0 / 0.8 = 7.5\,\text{cm}$$

Thus, apply irrigation when CPE (minus effective rainfall, if any) becomes equal to 7.5 cm. The daily evaporation can be recorded by using a pan evaporimeter.

3.4.2 Amount of Irrigation

The amount of water to be applied during irrigation is also known as *irrigation depth*. Correct amount of water needs to be applied in each irrigation event to improve crop water-use efficiency. Over- and under-irrigation may lead to low WUE and loss of crop yield and quality. The irrigation depth depends on the soil texture, crop species, rooting pattern, root zone depth, and method of irrigation. There are several ways to determine the irrigation depth.

3.4.2.1 Field Water Balance Method

The irrigation depth equals the cumulative water loss through evapotranspiration minus rainfall during the period, if any. A daily record of crop ET (actually determined or estimated) is maintained.

3.4.2.2 Field Capacity Deficit

If data on crop ET are not available, irrigation depth may be computed using data on field capacity deficit in the root zone. Irrigation is applied when the maximum permissible water depletion limit (MAD) is reached in the root zone. The irrigation depth (D_i) is calculated as follows:

$$\text{Net irrigation depth}\left(D_{\text{i}}\right) = \left(\text{PAWC} \times \text{MAD}\right) D_{\text{e}} \tag{3.2}$$

where PAWC is the plant available water capacity (m/m soil depth), MAD the maximum allowable water depletion limit (fraction), and D_e the root zone depth

Table 3.5 The average plant available water capacity (PAWC) of texturally different soils

Soil texture	PAWC (m/0.3 m)
Coarse sands	0.006–0.020
Fine sands	0.020–0.025
Loamy sand	0.028–0.030
Sandy loams	0.032–0.036
Fine sandy loam	0.038–0.050
Loam	0.046–0.050
Silt loams	0.050–0.064
Clay loam	0.046–0.050
Silt clay loams	0.046–0.050
Silty clay	0.038–0.043
Clay	0.030–0.038

(m). The average values of PAWC of some soils are shown in Table 3.5. Although the MAD limit depends on the crop and soil type, a value of 0.5 is applicable to most of the crops and soils. This method gives reasonably reliable estimate of net irrigation depth.

Once the net irrigation depth is known, the gross irrigation depth is calculated taking into consideration the irrigation efficiency of the system. The gross irrigation is always higher than the net irrigation depth because no irrigation system is 100% efficient. Gross amount is calculated from net amount of water as follows:

$$\text{Gross amount to be applied} = \text{Net amount} / \text{irrigation efficiency} \quad (3.3)$$

Numerical example: If plant available water capacity (PAWC) of soil is 160 mm/m, maximum permissible water depletion limit (P) is 0.5 (fraction) and root zone depth (D_e) is 0.4 m, and calculate the net irrigation depth (D_i mm) to replenish the moisture deficit. If the irrigation efficiency of the system is 70%, what will be the gross irrigation depth?

Solution:

According to Eq. 3.2,

$$\text{Net irrigation depth}\left(D_\text{i}\right) = \left(\text{PAWC} \times P\right) D_\text{e}$$

$$D_\text{i} = 160 \times 0.5 \times 0.4 = 32\,\text{mm}$$

Thus, 32 mm water is to be applied to the field as and when 32 mm of water has been depleted in the root zone.

$$\text{Irrigation efficiency of the system} = 70\%$$

From Eq. 3.3,

$$\text{Gross irrigation depth} = (32 \times 100) / 70 = 45.7\,\text{mm, i.e., } 46\,\text{mm}$$

Thus, to replenish the 32 mm water deficit in the root zone, 46 mm water must be applied through irrigation when irrigation efficiency is 70%. The gross irrigation may be decreased by using a more efficient irrigation system.

3.4.2.3 Use of Tensiometers

The tensiometers used to determine the time of irrigation can also be used to determine the irrigation depth (Sect. 3.4.1.2). Irrigation is stopped when the tensiometer installed at two-thirds depth of active root zone starts to show a rise in matric suction. It is an indirect way of knowing when to stop irrigation.

3.4.3 Methods of Irrigation

The third important component of irrigation scheduling is the method of irrigation. The objective of selecting a particular irrigation method is to:

- Maximize irrigation efficiency by minimizing nonproductive water losses during conveyance from the source to the field and achieving uniform water application in the field
- Minimize labor requirement, energy use, and cost of irrigation

Different methods of irrigation may be grouped into three categories (Fig. 3.1):

(a) *Surface methods*, wherein water is applied at the soil surface by gravity flow, viz. surface flooding, check basin and basin method, border strip method, and furrow method
(b) *Subsurface methods*, wherein water is applied below the surface, viz. use of open ditches, perforated pipes placed below the root zone, and mole drains
(c) *Micro-irrigation* (also called *pressurized irrigation*) *methods*, wherein water is applied under pressure, viz. sprinkler and drip irrigation

The choice of the method depends on several crop, soil, and climatic parameters, such as topography of the land, slope of the field, soil texture, infiltration rate, amount and quality of irrigation water available, type of the crop, climatic factors, and above all the cost involved and financial condition of the farmer.

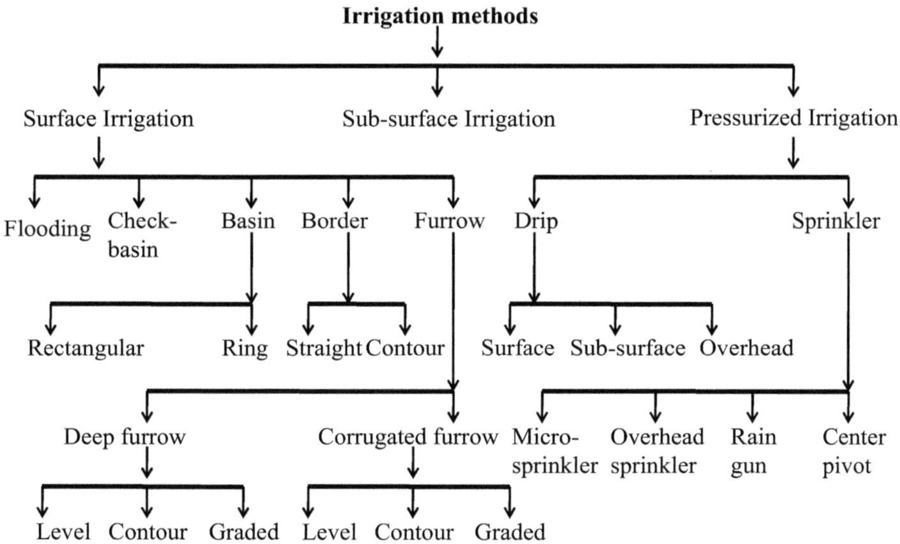

Fig. 3.1 Different methods of irrigation

3.4.3.1 Surface Irrigation

Surface irrigation consists of a broad class of irrigation methods (Fig. 3.1) in which water is distributed over the soil surface by gravity flow. Surface flooding is an ancient technique of water application in plain areas world over and is still commonly practiced for cereals, pulses, and oilseed crops. *Terrace irrigation*, the flood irrigation on terraced lands, is the oldest method of irrigation in hilly regions all over the world, including in Peru, Syria, China, and India. Around 95% of irrigated lands globally are irrigated by surface irrigation methods.

In surface flooding, the irrigation water is introduced onto the flat fields, level or graded furrows, or basins, using siphons, gated pipe, or turnout structures, and is allowed to advance across the field under gravity. Surface irrigation is best suited to flatlands and medium- to fine-textured soils which promote the lateral spread of water down the furrow row or across the basin. There are several variations of surface irrigation system:

- Wild flooding
- Border strip method
- Check basin method
- Furrow irrigation
- Raised-bed planting system
- Surge flow irrigation

Raised-bed planting system is a modification of ridge-furrow system. In this system, ridges are replaced by flatbeds which are planted to crops. Irrigation is

Table 3.6 Important features of some surface irrigation methods

Parameter	Wild flooding	Border strip irrigation	Check basin irrigation	Furrow irrigation
Mode of water application	Water diverted to the fields along the natural slopes without much control over it	Water diverted to individual field strips bounded by levees or borders; strips are narrow (10–20 m wide) and long (100–300 m) with gentle slope (0.05–0.5%)	Water diverted to small plots (i.e., checks) separated by levees, or to basins around one or more trees	Water applied in furrows (10–15 cm deep, 10–30 cm wide)
Irrigation efficiency	<30%; waste of water and nonuniform distribution	30–60%	60–75%	60–80%
Suitability	Cereals, fodder crops, pasture lands, etc.	Cereal crops in highly permeable soils	Small leveled plots and orchards	Row crops and vegetable crops

applied in furrows between the flatbeds. *Surge flow irrigation* is a method wherein water is applied intermittently in a series of on-and-off modes of constant or variable time spans. This system of irrigation has the potential to increase irrigation efficiency and save water. Important features of different surface irrigation methods are summarized in Table 3.6.

Surface irrigation methods, especially wild flooding, are comparatively less efficient than subsurface or pressurized irrigation systems. But with precise grading of the topography, high instantaneous flow rates, and relatively high levels of automation and management, the efficiency of these methods can be significantly improved.

3.4.3.2 Subsurface Irrigation

In subsurface irrigation, water is applied below the soil surface. Water reaches plant roots through capillary action. The subsurface irrigation is generally used in permeable soils and in soils having impermeable layer at or below 2 m depth or a high water table. Subsurface irrigation is not applicable in arid or semiarid regions where irrigation is often needed for crop germination. There are comparatively few regions globally where subsurface irrigation is applicable.

Water is regulated above the impervious layer using open ditches, deep furrows, mole drains, and perforated pipes placed below the root zone, etc. to maintain optimum moisture in the root zone through capillarity. Open ditches, usually 0.6–0.9 m deep and 0.3 m wide, are dug 45–90 m apart. Subsurface drainage system (open ditches or perforated pipes) used to lower the water table below the root zone during wet season may be used for subsurface irrigation during dry periods by feeding water into the drains and closing their outlets. In soils which allow free lateral movement of water, rapid capillary movement in the root zone soil, and very slow downward movement of water in the subsoil, water may be applied through a pipe distribution system placed below the soil surface. In soils having water table well below the surface, drip or trickle irrigation systems are placed beneath the soil surface, usually within the plant root system. Water table with good-quality water occurring within 1 m depth from the soil surface naturally serves as a subsurface irrigation system. The efficiency of subsurface irrigation is higher than surface irrigation.

Subsurface irrigation has the advantage of minimum evaporation loss, little field preparation, and less labor cost. Irrigation efficiency may be around 50–80%. Initial installation cost may be high. Improper flow rates in supply ditches, however, may result in waterlogging.

3.4.3.3 Micro-Irrigation

In micro-irrigation system, water is applied under pressure in the form of rain-like drops or fine sprays. The pressure is generated by mechanical pumps. Water can also be applied under gravity on sloping lands. Micro-irrigation systems were initially introduced to irrigate high-value cash crops and orchards, especially in areas where water was scarce and costly and topography was irregular. But now, micro-irrigation systems are used to irrigate a wide range of crops and orchards and under diverse conditions. Micro-irrigation is useful in undulating terrain, rolling topography, hilly areas, barren land, and areas having shallow soils, where surface irrigation is not possible or feasible. Water distribution uniformity is higher with micro-irrigation (drip >90% and sprinkler irrigation 60–80%) than with surface irrigation (50–60%).

The micro-irrigation systems can be operated manually or automated by using rain sensors or soil moisture sensors. These sensors automatically switch on and switch off the drip and sprinkler irrigation devices once a prefixed amount of water has been applied, thereby reducing labor requirement and operational cost and avoiding over-irrigation of fields and thus saving irrigation water.

Micro-irrigation system may broadly be grouped into two categories:

1. Drip irrigation system
2. Sprinkler irrigation system

In addition to these two categories, there are other types of micro-irrigation systems also, such as *bubblers* and *micro-sprayers*. Bubblers apply water to individual plants in an umbrella pattern. The micro-spray system uses jets, foggers, or misters, called *spitters*, to apply water. Water is ejected as fine jets that fan out from a series of nozzles. The drip and sprinkler irrigation systems have distinct differences in the water flow rate (i.e., water application rate, liters/minute), operating pressures, and wetted area.

I. *Drip Irrigation:* Drip irrigation system is also known as *trickle irrigation system*. In this system, water is applied directly to the root zone in the form of drops or by a micro-jet on soil surface or subsurface through drippers or emitters. The emitters are generally placed on soil surface, but they may also be placed beneath the soil surface in the root zone. The flow rates are kept low (2–20 L/h), i.e., lower than the infiltration rate of soil. The water is applied daily or at very short intervals of 1–3 days. Drip irrigation keeps the root zone always moist for the plants to transpire at potential/near-potential rates.

Drip irrigation system has several variants:

(i) *Surface drip irrigation*: used to irrigate annual and perennial plants.
(ii) *Sub-surface drip irrigation*: crops are irrigated through 15–25 cm deep buried plastic tubes, containing embedded emitters located at regular spacing.
(iii) *Family drip*: also called *gravity-fed drip irrigation*; suitable for irrigated small areas (500–1000 m^2) like kitchen gardens, peri-urban agriculture, or demonstration plots; cheap and easy to install and operate.
(iv) *Online drip*: emitters or drippers are fixed externally on the laterals at designed spacing, and the spacing can easily be changed to accommodate increased root zone; mostly used in orchards, vineyards, artificial landscapes, and nurseries.
(v) *In-line drip*: drippers are fixed in the lateral tube at designed spacing at the time of manufacturing; dripper spacing cannot be changed; suited to row crops like cotton, sugarcane, groundnut, vegetables, and flowering crops.

Drip irrigation can also be operated on windy days and during various land operations. Drip irrigation differs from other irrigation systems in that it wets only the root zone, whereas surface or sprinkler irrigation systems involve wetting the whole soil profile. Also, the frequency of water application with drip irrigation system is very high compared to other irrigation systems.

Drip irrigation is adaptable to any slope where farming is possible. It is suitable for most soils. On clay soils, water must be applied slowly to avoid surface water ponding and runoff. On sandy soils, higher emitter discharge rates will be needed to ensure adequate lateral wetting of the soil. It is most suited to row crops (vegetables, soft fruit), tree, and vine crops where one or more emitters can be provided for each plant. Drip irrigation is usually considered for high-value crops because of the high capital costs of installing a drip system.

Advantages of drip irrigation system:

- Irrigation scheduling can be managed precisely to meet crop demands. Water reaches directly to the plant roots. It holds promise of increased yield and quality.
- Irrigation efficiency is very high. A well-designed drip irrigation system loses practically no water to runoff, deep percolation, or evaporation. The saved water can be utilized for irrigating more area. It also makes the system particularly suitable to water-scarce areas or where water is expensive.
- Drip irrigation ensures quality produce. It reduces water contact with crop leaves, stems, and fruit, thereby making conditions unfavorable/less favorable for the onset of diseases. The area near the plants remains dry, so bacterial growth is limited.

- Weed growth is low because it supplies water directly to the root zone and does not wet the entire soil surface.
- Fertigation and chemigation are possible with drip irrigation. It increases nutrient-use efficiency and decreases labor cost. Since only the root zone is irrigated, leaching losses of N are reduced and use efficiency of applied fertilizer N is enhanced. It economizes on the use of agrochemicals (insecticides, pesticides, fungicides, herbicides, etc.).
- Drip irrigation is feasible irrespective of topography, soil texture, and configuration of fields. Expansive land leveling or terracing is not required, and there is no loss of cultivable area due to channel construction. Drip irrigation can work well in areas where other irrigation systems perform poorly, as in highly permeable soils and soils prone to puddling or runoff.
- Being a permanent system[2], drip irrigation system can be easily automated.

Limitations of drip irrigation system:
There are some limitations of drip irrigation:

- Drip irrigation systems are expensive because of high capital investment ($1250–$3000 or more per ha). Annual maintenance cost is also high.
- Plugging of emitters by silt or other soil particles, algal growth, or chemical (fertilizers, pesticides, etc.) deposits is a common problem. All emitters have very small openings (0.2–2.0 mm in diameter), and these can become easily blocked. Leaking of emitters is also frequently observed. Many times, problem may be complex and the solution requires an experienced engineer.
- Surface-placed drip irrigation system may interfere with farm operations or may get damaged during farm operations (cultivation, weeding, etc.). The "subsurface drip irrigation" with drip system buried below the soil surface avoids such problems.

II. *Sprinkler Irrigation:* In sprinkler irrigation system, water is sprayed over or under the crop canopy in the form of fine droplets similar to rain or fine sprays. The water flow rate should not exceed the infiltration rate of soil. The flow rate can be regulated by changing the size of sprinkler nozzles and the operating pressure.

The sprinkler system consists of a main pipeline, joined to perpendicular pipelines, called *the laterals*, with rotating nozzles, i.e., the *sprinklers*. Water passes through these pipelines under pressure and gets sprinkled over the crop through rotating nozzles in the form of rain-like droplets. The sprinkler heads are available in different designs depending on the mode and pattern of water sprays, such as rotor-type sprinklers (impact-driven and gear-driven rotors); self-propelled irrigation systems, e.g., center-pivot sprinklers; linear-move sprinklers; traveling big gun system; and wheel-roll system. These sprinklers spray water over the crop canopy.

Depending on the design, operating system, portability (portable, semiportable, semipermanent, and permanent system), and spray pattern (rotating head or

[2] A system that remains in place during more than one cropping season.

revolving sprinkler system and perforated pipe system), there are several variants of sprinklers:

(i) Center pivot
(ii) Towable pivot
(iii) Rain gun
(iv) Impact sprinkler
(v) Pop-up sprinkler
(vi) Linear-move sprinkler
(vii) Micro-sprinkler

Sprinkler irrigation can be easily adapted regardless of topography and soil type for irrigating almost all types of crops, viz. field crops, vegetables, floriculture, and orchards. Due to the high capital investment, self-propelled irrigation systems (center pivots, linear moves, traveling big guns, and side-roll systems) are used on high-value crops such as potatoes and vegetables. Sprinkler irrigation is a common practice nowadays. The top ten countries using sprinkler irrigation are the United Sates, Russia, China, India, France, Brazil, Italy, Spain, Saudi Arabia, and Ukraine. These countries together constitute nearly 75% of total sprinkler-irrigated area.

Advantages of sprinkler irrigation system:

- Expansive land leveling or terracing is not required.
- No loss of cultivable area due to channel construction.
- Suitable for almost all types of soils and crops.
- Saves irrigation water due to its high irrigation efficiency.
- Protects the crops from extreme frost or temperature hazards.
- Maintains soil physical conditions (i.e., soil structure) and increases soil productivity.
- Labor requirement is low as compared to traditional surface irrigation methods.

Limitations of sprinkler irrigation system:

- High initial capital and operational costs.
- Winds lower water distribution and application efficiency and, coupled with high temperature, may enhance evaporation losses.
- Unavoidable wetting of foliage in field crops may result in increased sensitivity to diseases. It may affect the quality of produce.
- Under high-temperature conditions (≥35 °C), irrigation with saline water (>7 millimhos/cm) causes leaf burning.
- Salts, debris, and sediments in irrigation water can cause clogging of sprinkler nozzles.
- It interferes with farm operations.

<u>Salient features of different micro-irrigation systems:</u>

Salient features of drip, sprinkler, and micro-sprinkler irrigation systems are summarized in Table 3.7.

Table 3.7 Comparison among drip, sprinkler, and micro-sprinkler irrigation systems

S. No.	Drip irrigation	Sprinkler irrigation	Micro-sprinkler irrigation
1.	Water is applied daily at the root zone in the form of drops	Water is applied at frequent intervals in the form of fine spray	Water is applied daily or at frequent intervals in the form of spray over the crop canopy
2.	Operating pressure is low (≤1.5 kg/cm²)	Operating pressure is high (>2.5 kg/cm²)	Operating pressure is low to medium (1–3 kg/cm²)
3.	Water flow rates are low (2–20 L/h)	Water flow rates are high	Water flow rates are higher than drip and lower than sprinkler irrigation (40–75 l/h)
4.	Water is applied directly in the root zone; irrigation covers only a small fraction of the field	Irrigation covers the entire field	Irrigation may cover localized or full surface area
5.	Evaporation, deep percolation, and runoff losses during irrigation are eliminated	Evaporation losses during irrigation are high; deep percolation and runoff losses may also be observed	Evaporation losses during irrigation are high; deep percolation and runoff losses are minimal
6.	Irrigation efficiency is very high, as high as 95%	Irrigation efficiency is 60–85%	Irrigation efficiency is around 80–90%
7.	Water distribution efficiency (uniformity in water application) is very high, i.e., 90–95%	Water distribution efficiency is 60–80%	Water distribution efficiency is 80–90%
8.	Water application not affected by winds	Winds cause drift in water sprays, lowering water distribution efficiency	Water distribution is affected by winds
9.	Fertigation is feasible	Fertigation is not feasible	Fertigation is not feasible
10.	Weed growth is low	Weed growth is high	Weed growth is high
	Suitable for high-value row crops, vegetable crops, trees, orchards, vineyards, and for protected agriculture	Suitable for most row, field (cereal and fodder crops), and tree crops, and pasturelands	Suitable for seasonal and high-value crops; particularly suited for shallow-rooted crops like garlic and onion

Advantages of micro-irrigation system:

Micro-irrigation system is better than surface irrigation system in several ways:

- *Saving of irrigation water*: Micro-irrigation system can save 45–60% of irrigation water as compared to surface irrigation.
- *High level of uniformity in water application*: The water application uniformity is much higher with micro-irrigation than surface irrigation.
- *Fertigation and chemigation possible*: Fertigation with micro-irrigation, which is not possible with surface irrigation system, improves fertilizer-use efficiency, saves fertilizer, and reduces total cost by 25–50%. Similar benefits are achieved with the use of herbicides, insecticides, and fungicides with micro-irrigation system.

- *Reduces weeds and diseases*: Due to limited area irrigated, weed growth is inhibited with micro-irrigation, especially drip irrigation. It leads to reduction in insect-pests and diseases in plants because weeds are habitats of diseases and insect-pests.
- *Increases tolerance of plants to soil salinity*: Micro-irrigation (especially drip irrigation) maintains a low concentration of salts in the root zone by keeping soil moisture continuously at a high level. The crop plants, therefore, can tolerate irrigation with saline water in a better way with micro-irrigation.
- *Makes cultivation successful in problem soils*: With micro-irrigation, crops can be successfully grown in problem soils, viz. soils having very high or very low infiltration rates, salt-affected soils, and in areas having poor-quality water which could not be utilized with other irrigation methods.
- *Suitable to various topography and soil types:* Land leveling, etc. are not required.
- *Automation is possible with micro-irrigation*: The system may be partially or fully automated. The system may remain operative any time of the day and for any duration.
- *Micro-irrigation is cost effective:* Automated micro-irrigation system reduces labor requirement and cost.
- *Improves quality and yield of crops*: Slow, regular, and uniform application of water and nutrients with micro-irrigation improves quality and yield of crops. Crops mature early. According to a report of the National Mission on Micro Irrigation, India (2014), micro-irrigation saved irrigation water by 20–48%, energy by 10–17%, fertilizers by 11–19%, and labor cost by 30–40% and increased crop production by 20–38%, fruit production by 42%, and vegetable production by 53% with average reduction of irrigation cost by 32% compared to conventional (surface) irrigation methods. With the help of micro-irrigation, 519.43 ha of degraded land was brought under cultivation.

The micro-irrigation has seen fast development over the past three decades and has now become the standard for efficient irrigation practices for water conservation and optimal plant responses. Being a relatively new technology, micro-irrigation currently is in use on less than 1% of lands worldwide, and around 5% of the irrigated area in the United States. High initial capital cost is also responsible for its slow adoption. The technology, however, has great agricultural potential in the foreseeable future due to increasing demand for otherwise limiting water resource and increasing concerns about environmental consequences of irrigation.

3.5 Irrigation Efficiency

Irrigation efficiency (IE) indicates how efficiently water is applied to the field. Water engineers define IE as "the ratio of the net amount of water stored in the root zone to the amount of water diverted from the source." Crop physiologists/

agronomists define IE as "the ratio of the amount of water consumed by the crop to the amount of water supplied through irrigation."

Irrigation efficiency (IE) has four components: water conveyance efficiency (WCE), water application efficiency (WAE), water storage efficiency (WSE), and water distribution efficiency (WDE).

- *Water conveyance efficiency* (WCE) refers to "the ratio of water reached at the field to the water diverted from the source." A portion of water may be lost during conveyance from the water source to the field/farm through seepage, percolation, runoff, evaporation, consumption of water by the unwanted vegetation growing in water channels, etc. Such losses depend largely on the quality of channels, e.g., earthen or concrete channels, lined or unlined channels, and PVC/ metal pipes.
- *Water application efficiency* (WAE) refers to "the ratio of water stored in the root zone to the water applied to the field." The WAE depends on the layout of the field/farm and the method of irrigation. Good layout and land leveling improve WAE. Water application to the uneven field, runoff, deep percolation from the field, etc. lower WAE.
- *Water storage efficiency* (WSE) refers to "the ratio of the irrigation water stored in the root zone to the water storage needed in the root zone (i.e., root zone water deficit) before irrigation." Some people call it *water requirement efficiency* (WRE) (Koech and Langat, 2018).
- *Water distribution efficiency* (WDE) indicates how uniformly the water is stored in the root zone. The WDE is computed by using the following equation:

$$\text{WDE} = 1 - \left(N / D\right) \tag{3.4}$$

where N is the average numerical deviation in depth of water stored at different locations in the field from the average water storage depth of field following irrigation, and D the average water storage depth of the field.

The product of these four components gives the irrigation efficiency (IE):

$$\text{IE} = \text{WCE} \times \text{WAE} \times \text{WSE} \times \text{WDE} \tag{3.5}$$

3.5.1 Improving Irrigation Efficiency

The improvement in on-farm irrigation efficiency saves water which can be diverted to irrigate additional area, conserves water resources, improves crop production, increases net returns, and provides long-term sustainability of agriculture production system. Any option/technology that decreases water conveyance losses, increases water application efficiency, and gives highest water distribution efficiency improves the irrigation efficiency. Different ways of improving IE are briefly described below:

- *Decrease water conveyance losses in field water channels*: Conveyance losses in an open channel can be significantly reduced by ditch lining, ditch realignment, or installing a closed pipeline and removing unwanted vegetation (weeds, shrubs, etc.) from the unlined portion of water channels.
- *Improve water distribution efficiency* (WDE) to reduce deep percolation: For surface irrigation systems, it can be achieved through land leveling with small uniform gradient, relatively fast water application rate, surge irrigation, furrow firming, etc. For sprinkler systems, options include changing sprinkler types, renozzling the system, or changing nozzle spacing to improve the overlap between heads. The WDE is different with different irrigation methods. WDE is higher with drip (>90%) and sprinkler irrigation (60–80%) than with surface irrigation (50–60%).
- Try to *match irrigation amount with the WHC of soil profile* so as to reduce percolation and runoff losses.
- *Select a more efficient irrigation system.* Different irrigation systems have different irrigation efficiencies (Table 3.8).

3.6 Fertigation

Fertigation is a process that combines fertilizer application and irrigation. *Fertigation* may be defined as "the practice of applying fertilizers to crops along with irrigation." Other chemicals, like fungicides, insecticides, and herbicides, may also be applied along with irrigation, and the process is called *chemigation.* Fertigation is done using drip irrigation or micro-sprinkler systems and is most commonly used by commercial growers. The simplest type of fertigation system consists of a tank with a pump, distribution pipes, capillaries, and a dripper pen. In order to avoid the potential risk of contamination in the potable (drinking) water supply, a backflow prevention device is required for most fertigation systems.

Table 3.8 Irrigation efficiency of different irrigation systems

Method of irrigation	Irrigation efficiency (%)
Surface irrigation	30–70
Subsurface irrigation	50–80
Sprinkler irrigation	50–90
Drip irrigation	75–95
Micro-irrigation	≥95

Source: Gilley and Watts (1977), Sivanappan (1994), Seckler et al. (2003)

Table 3.9 Commonly used fertilizers in fertigation in India

Fertilizer	N-P_2O_5-K_2O content (%)	Solubility (g/l) at 20 °C
Ammonium nitrate	34-0-0	1830
Ammonium sulfate	21-0-0	760
Urea	46-0-0	1100
Monoammonium phosphate	12-61-0	282
Diammonium phosphate	18-46-0	575
Potassium chloride	0-0-60	347
Potassium nitrate	13-0-44	316
Potassium sulfate	0-0-50	110
Monopotassium phosphate	0-52-34	230
Phosphoric acid	0-52-0	457
Multi-K	13-0-46	–
Polyfeed	$N = 11–20$ $P_2O_5 = 8–42$ $K_2O = 11–30$	–

Source: Tamil Nadu Agricultural University, Agritech Portal, India

3.6.1 Commonly Used Fertilizers in Fertigation

Two types of fertilizers are used in fertigation: (i) solid fertilizers that can properly dissolve in water and (ii) liquid fertilizers. The commonly used fertilizers in fertigation in India are listed in Table 3.9.

Fertilizers like monoammonium phosphate (N and P), polyfeed (N, P, and K), multi-K (N and K), and potassium sulfate (K and S) are highly suitable for fertigation. Micronutrients such as Fe, Mn, Zn, Cu, B, and Mo can also be supplied along with these fertilizers.

The parameters considered for selecting fertilizers for fertigation include solubility, compatibility, acidity, and salinity (osmotic pressure).

- Fertilizers must have quick solubility, forming a clear solution in water, and should not precipitate out of solution. Fertilizers like ammonium nitrate, potassium nitrate, urea, and ammonium phosphate are quick water-soluble fertilizers. Fertilizers like monoammonium phosphate, urea phosphate, and phosphoric acid form precipitates when added in higher concentrations in hard water or at low temperatures.
- Fertilizers must be compatible and should not form precipitates upon mixing.
- The fertilizer solutions should not be too acidic or too alkaline to avoid corrosion of equipment and precipitation of salts. Acid solutions have high corrosivity, while alkaline liquids represent the risk of precipitation.
- The nature and concentration of fertilizers used in fertigation should not decrease the osmotic potential of water so as to retard water absorption by plant roots.

Table 3.10 Fertilizer-use efficiency through fertigation

Nutrient	Fertilizer-use efficiency (%)	
	Soil application	Fertigation
N	30–50	95
P	20	45
K	50	80

Source: Tamil Nadu Agricultural University, Agritech Portal, India

Table 3.11 Water saving and crop yields with drip irrigation and drip fertigation

Crops	Water saving (%)	Yield (t/ha)		
		Conventional	Drip irrigation	Drip fertigation
Banana	35	26	30	37
Sugarcane	29	120	160	207
Tomato	32	45	56	65

Source: Tamil Nadu Agricultural University, Agritech Portal, India

3.6.2 Advantages of Fertigation

Fertigation has some distinct advantages over the broadcast or band placement of fertilizers. Some advantages are listed below:

- *Higher nutrient and water-use efficiency*: Precise and uniform application of nutrients and water in the active root zone improves nutrient and water absorption by the plant roots. Nutrient absorption starts immediately after application of fertilizers in liquid form. It also promotes rapid root growth. Nutrient- and water-use efficiencies are significantly improved (Table 3.10).
- *Higher yields and better quality crops:* Nutrients are precisely applied in the active root zone according to the nutritional requirements of the crop. A frequent supply of water and nutrients reduces fluctuation of nutrient concentration in soil and avoids water and nutrient stress at any physiological crop stage. The result is saving of water and nutrients and higher crop yields of better quality (Table 3.11). Fertigation may give fertilizer-use efficiency as high as 80–90%, nutrient saving at least by 25%, and yield increase by 25–50%. Fertigation with drip irrigation improves the quality of fruits and flowers while with micro-sprinkler irrigation improves the quality of leafy vegetables.
- *Efficient application of microelements*: Micronutrients (Fe, Mn, Zn, Cu, B, Mo) are expensive and are required in small quantities. They are conveniently and efficiently applied through fertigation.
- *Economy of time and resources*: Micro-dosing of fertilizers in each application, reduction in leaching losses of fertilizers, and combining two processes of fertilizer application and irrigation into one economize time and resources (nutrients and water) and decrease production costs. Application of nutrients only to the wetted soil volume reduces loss of nutrients by leaching or soil fixation and increases fertilizer-use efficiency.

- *Reduction of groundwater pollution*: The exact dosages optimize fertilizer application, reduce nutrient leaching, and reduce groundwater contamination hazards.
- *Reduction of soil erosion*: Low water application rates through fertigation reduce soil erosion hazards.
- *Reduction of biotic stresses*: Crop foliage remains dry particularly with fertigation using drip irrigation. It reduces incidences of pests and diseases.
- *Convenience of handling*: Fertigation with fertilizer solutions is more practical and convenient than the broadcast or band placement application of solid or granular fertilizers.

3.6.3 Limitations of Fertigation

- Subjective application of fertilizers is very difficult in gardens or farms that have different types of plants. Fertigation supplies nutrients and water uniformly to all plants in one go.
- Concentration of the solution may decrease with time during application as the fertilizer dissolves. It may lead to poor nutrient placement.
- Possible pressure loss in the main irrigation line may lead to problem of nonuniform nutrient placement.
- Saline or unclean/muddy water or precipitates of noncompatible fertilizers may clog the drippers and disrupt application uniformity. Irrigation water must be clean and nonsaline.
- If care is not taken, fertigation may contaminate the potable water source.

3.7 Irrigation Management in Saline Soils

Salt-affected soils (saline and alkali soils) adversely affect crop productivity by accumulating salts in the root zone beyond levels critical for plant growth and also deteriorating soil (physical) properties. Several factors are responsible for salinization including over-irrigation, seepage, impeded drainage, waterlogging, rapid rise of water table causing secondary salination, etc. With proper irrigation management, (a) saline soils can be made fertile and (b) salinization of normal soils can be prevented. The irrigation water is so managed as to leach the excess salts from the root zone, keeping salt concentration below the critical values for the crop plants. The principle is to apply irrigations frequently with water adequate to meet the leaching requirement of the field.

Leaching requirement (LR) refers to “the water needed to remove excess salts from the root zone that decrease the crop yields.” The US Salinity Laboratory (1954) defined LR as “the fraction of the irrigation water that must pass through the root zone in order to prevent average soil salinity from rising above some specifiable level." The LR depends on the salt concentration of the irrigation water (EC_{iw}), the salt concentration in the root zone, and the salt tolerance of the crop to be cultivated.

The fraction of the applied irrigation water that drains beyond the root zone is called the *leaching fraction* (LF). Under steady-state conditions, the LF is approximated by the following equation:

$$LF = D_{dw} / D_{iw} = EC_{iw} / EC_{dw} \tag{3.6}$$

where D_{dw} is the depth of drainage water (mm), D_{iw} the depth of irrigation water (mm), EC_{iw} the electrical conductivity of the irrigation water (dS/m), and EC_{dw} the electrical conductivity of the drainage water (dS/m). The EC_{dw} indicates the salt concentration of the root zone. Thus, the LF indicates the degree to which salts are leached from the root zone.

The LR represents the minimum LF that could be allowed without soil salinity becoming excessive for optimal plant growth. The permissible limit of yield decrement to calculate LR is 10% from the maximum. Thus, the minimum value of LF (i.e., LR) would be given when the maximum permissible salinity level of EC_{dw} (i.e., D_{dw}) is used in Eq. 3.6, i.e.,

$$LR = EC_{iw} / EC_{dw} \tag{3.7}$$

Equation 3.7 considers only the salt tolerance of the crop and the salt concentration of the irrigation water and is applicable under steady-state conditions.

The *steady-state condition* is one when water content and salt concentration remain constant over time at a given soil depth. Under most field situations, however, steady-state conditions do not exist. Several factors keep disturbing steady-state condition, including water extraction by roots, water replenishment by irrigation, crop type, variation in irrigation water quality, and alteration in irrigation management. In addition, LR is influenced by numerous factors, including irrigation nonuniformity, mineral precipitation-dissolution reactions, transient root water uptake distributions, preferential flow, climate, runoff, extraction of shallow groundwater, and leaching from effective precipitation, and the assumption of steady-state conditions. These factors are not taken into consideration while calculating LR. Hence, the reliability of Eq. 3.7 in calculating LR becomes questionable (Corwin et al. 2012). Nevertheless, the LR model is still widely used for irrigation management in salt-affected soils.

It may be noted that for the efficient utilization of irrigation water, an accurate and reliable method of calculating the LR is important. Underestimation of LR would result in salt accumulation in the root zone and yield reduction, and overestimation would result in excessive water utilization and nutrient removal producing detrimental environmental impacts on groundwater or degraded drainage waters. Water depth during each irrigation equals the sum of crop water requirement and leaching requirement. Water can be applied using a suitable irrigation method.

3.8 Irrigation Management in Highly Permeable Soils

Irrigation scheduling in highly permeable soils must focus on two things: timing and amount of irrigation applied. Application of too much water (deep and frequent irrigations) may lower water-use efficiency, increase chances of leaching of plant nutrients and other agrochemicals resulting in pollution of groundwater resource, and increase the cost of irrigation (pumping cost, labor cost, etc.). Under-irrigation (less frequent and shallow irrigations) or delayed irrigation until water stress becomes evident on plants may result in economic yield loss.

Micro-irrigation systems are most suited to highly permeable soils. If coupled with mulches, the water-use efficiency is further increased. Soils having infiltration rates >100 mm/h are generally not suited to surface irrigation systems. If at all surface irrigation has to be used, for uniform water application, the size of fields must be conveniently small and the irrigation application rates must be large, and for minimizing conveyance water losses, the length of conveyance channels must be short and the channels must be lined.

3.9 Management of Saline Irrigation Water

Irrigation waters always contain dissolved salts in them. The salt concentrations vary with climate (e.g., waters in arid and semiarid regions are more saline than in humid regions) and source of water (snowmelt and rainwaters are less saline than groundwaters in arid and semiarid regions). The commonly found ions in irrigation waters include calcium, magnesium, sodium, potassium, chloride, bicarbonate, carbonate, sulfate, and others. Some effluent waters may contain higher boron concentrations. In higher concentrations, these ions are toxic to crop plants. Scarce water conditions compel to use saline waters for irrigation.

The source of saline waters may be the brackish water aquifers or drainage water from waterlogged areas. The management of saline waters requires specific considerations depending on the level of salinity, crop type, and soil type. Following considerations are important for using saline irrigation waters:

- Light and frequent irrigations should be avoided as they increase salt concentrations in the topsoil/root zone.
- Avoid sprinkler irrigation or other irrigation methods that produce fine droplets or mist and wet the crop foliage. Salts coming in direct contact with leaves may cause leaf burning. Surface drip irrigation is most suited for saline waters as it does not wet plant foliage and also reduces salinity hazards to plants by maintaining continuously moist regime in the root zone. Other irrigation methods suited to saline waters include subsurface drip irrigation, low-energy precision application (LEPA) irrigation, and furrow irrigation. Clogging of drip emitters or filters, etc. due to precipitation of salts like calcium and magnesium carbonates is a common problem associated with the use of saline waters. To avoid clogging

Table 3.12 Tolerance of some crops to salinity in irrigation water

	Threshold EC in irrigation water (mmhos/cm or dS/m)	
Crop	10% Yield reduction	50% Yield reduction
Wheat	4.0	8.7
Barley	5.0	12.0
Corn	1.1	3.9
Sorghum	2.7	7.2
Soybean	3.3	5.0
Cotton	5.1	12.0
Alfalfa	1.3	5.9
Bermudagrass	4.6	9.8

Adapted from Rhoades et al. (1992) and Fipps (2003)

of emitters, etc., regular or occasional acid injections in water are needed. The occasional acid injections dissolve the precipitates, and the regular addition of acids prevents chemical precipitation.

- Apply saline water more than the actual crop water requirement to meet the leaching requirement of soil. The excess water is known as the "leaching fraction" (see Eq. 3.6).
- In regions where few water sources differing in salinity levels are available, there are three options of using these waters for irrigation: (i) blending of normal (low EC) and saline waters before irrigation (Kan and Rapaport-Rom, 2012), (ii) substitution of saline water with normal water in sequence (Sang et al. 2020), and (iii) selective application of unmixed waters to crops or group of crops depending on their salinity tolerance. Such options depend on the quality of available waters, cropping patterns, soil properties, irrigation systems, climate, etc.
- If possible, minimize irrigation applications by selecting salt-tolerant crops (Table 3.12).

3.10 Question Bank

3.10.1 Short Questions:

(i) What is irrigation? What are its objectives?
(ii) Briefly describe the concept of irrigation.
(iii) What do you understand by irrigation scheduling? How is it important in irrigation management?
(iv) Is "irrigation scheduling" and "irrigation management" the same?
(v) How will you schedule irrigation based on the crop appearance? What are the limitations of this method?
(vi) What is the principle of "plant canopy temperature" as an index of irrigation scheduling?

(vii) Briefly explain the term "maximum allowable/permissible water depletion limit" (MAD). What is FAO's recommendation about this limit for different crops?
(viii) How will you schedule irrigation for wheat using tensiometers?
(ix) What is the principle of using "IW/CPE ratio" as an index of irrigation scheduling?
(x) How is irrigation depth calculated in the field? Compute the net irrigation depth (mm) if plant available water capacity of soil is 160 mm/m, maximum permissible water depletion limit is 0.5, and root zone depth is 0.4 m.
(xi) What are different methods of irrigation? Compare irrigation efficiency among wild flooding, sprinkler, and drip irrigation systems.
(xii) Briefly explain subsurface system of irrigation. What are the important limitations of this method of irrigation?
(xiii) What are the advantages of drip irrigation over the sprinkler irrigation?
(xiv) Define irrigation efficiency. What are its different components?
(xv) In what way irrigation efficiency can be increased? Compare irrigation efficiency of different irrigation systems, viz. surface irrigation, subsurface irrigation, sprinkler irrigation, micro-sprinkler irrigation, and drip irrigation.
(xvi) What is fertigation? What important parameters are considered while selecting fertilizers for fertigation? What is its significance in intensive agriculture?
(xvii) What are the advantages and limitations of fertigation?
(xviii) Explain "leaching requirement" and "leaching fraction." What is their significance in irrigation management?
(xix) What are the important considerations while using saline waters for irrigation?

3.10.2 *Briefly Explain Why?*

(i) Irrigation is essential even in high-rainfall areas.
(ii) Proper irrigation scheduling is a must to improve crop water-use efficiency.
(iii) Plant canopy temperature can be used for irrigation scheduling.
(iv) Irrigation is applied as soon as maximum permissible depletion limit is reached in soil.
(v) Better apply irrigation before maximum permissible depletion limit is reached.
(vi) Keeping root zone moist improves crop biomass production.
(vii) Drip irrigation system gives higher irrigation efficiency than sprinkler irrigation system.
(viii) Fertigation gives higher nutrient- and water-use efficiency.
(ix) Fertigation produces better quality crops.
(x) While irrigating with saline water, some amount of water must pass through the root zone.
(xi) Avoid sprinkler irrigation if the irrigation water is saline.

3.10.3 Fill in the Blanks:

(i) Drip irrigation with water carrying chemical fertilizers dissolved in it is called ……….
(ii) The thumb rule for most of the crops is to irrigate when half of the ………. in the root zone is depleted.
(iii) For better plant growth, it is better to apply irrigation before ………. is reached.
(iv) The …………. refers to the difference between the canopy temperature of water-stressed and unstressed crop.
(v) The difference in canopy temperature of water-stressed and unstressed crop may be as high as …… or even more.
(vi) The water content below which crop starts experiencing water stress is known as ……………….
(vii) ……….. is the commonly used equipment for measuring soil matric potential.
(viii) The range of irrigation efficiencies with surface irrigation, subsurface irrigation, and sprinkler and drip irrigation is ……., ..., …….., and ……. percent, respectively.
(ix) Fertigation may increase N-use efficiency from ….. percent (with soil application) to …….. percent.
(x) Fertigation may increase P-use efficiency from ….. percent (with soil application) to …….. percent.
(xi) Fertigation may increase K-use efficiency from ….. percent (with soil application) to …….. percent.

[**Key:** (i) fertigation; (ii) PAWC; (iii) maximum allowable depletion limit (MAD); (iv) stress degree-days, i.e., SDD; (v) 8 °C (vi) maximum allowable/permissible water depletion limit (MAD); (vii) tensiometer; (viii) 30–70, 50–80, 50–90, and 75–95; (ix) 50, 95; (x) 20, 45; (xi) 50, 80]

3.10.4 State Whether the Following Statements Are True (T) or False (F):

(i) Fertigation improves water- and nutrient-use efficiency.
(ii) Irrigation is also used to decrease soil penetration resistance to facilitate root growth.
(iii) The *qanats*, developed in ancient Persia (present-day Iran) about 6000 BC, are among the oldest known irrigation methods in use today.
(iv) The thumb rule for most of the crops is to irrigate when half of the PAWC in the root zone is depleted.

(v) Vegetative stage of wheat is more sensitive to water stress than the flowering stage.
(vi) Plant water status is a better index of irrigation scheduling than soil water status.
(vii) The readily available water content is the product of PAWC and the maximum depletion limit P.
(viii) The ease of water absorption by the plant roots declines progressively as the water potential decreases from −0.1 to −15.0 bar.
(ix) Plants must transpire water at a potential rate for accumulating maximum biomass.
(x) Micro-sprinklers give higher irrigation efficiency than drip irrigation.
(xi) Fertigation is feasible with drip irrigation but not with sprinkler irrigation.
(xii) Micro-sprinklers are ideally suited for fertigation.
(xiii) Only liquid fertilizers are used for fertigation.
(xiv) Solid fertilizers can also be used for fertigation.
(xv) The pH is an important characteristic of fertilizer solution used in fertigation.
(xvi) Fertigation may increase N-use efficiency by as high as three times over the one with soil application.
(xvii) Overestimation of LR would generally result in salt accumulation in the root zone.
(xviii) Deep and frequent irrigations in sandy soils increase crop WUE.
(xix) Salinity tolerance of wheat is higher than corn.
(xx) Avoid light and frequent irrigations with saline water especially in clayey soils.

[**Key:** (i) T, (ii) T, (iii) F, (iv) T, (v) F, (vi) F, (vii) T, (viii) T, (ix) T, (x) F, (xi) T, (xii) F, (xiii) F, (xiv) T, (xv) T, (xvi) T, (xvii) F, (xviii) F, (xix) F, (xx) T]

3.10.5 Multiple-Choice Questions:

(i) Time required to apply 10 cm of water to an area of 1.0 ha at a flow rate of 4.0 m^3/min is

(a) 200 min
(b) 250 min
(c) 300 min
(d) 350 min

(ii) One of the following is not a component of irrigation scheduling:

(a) Why to irrigate?
(b) When to irrigate?
(c) How much to irrigate?
(d) How to irrigate?

(iii) If PAWC is 150 mm/m and maximum allowable water depletion limit is 0.5, then readily available water content would be

(a) 30 mm/m soil depth
(b) 45 mm/m soil depth
(c) 60 mm/m soil depth
(d) 75 mm/m soil depth

(iv) Based on IW/CPE ratio of 0.8, to apply 6 cm water, apply irrigation when CPE equals

(a) 6.5 cm
(b) 7.0 cm
(c) 7.5 cm
(d) 8.0 cm

(v) If plant available water capacity of soil is 160 mm/m, maximum permissible water depletion limit is 0.5, and root zone depth (D_e) is 0.4 m, the net irrigation depth (D_i, mm) would be

(a) 32 mm
(b) 64 mm
(c) 128 mm
(d) 200 mm

(vi) Highest irrigation efficiency is obtained with

(a) Furrow irrigation
(b) Subsurface irrigation
(c) Micro-sprinkler irrigation
(d) Drip irrigation

(vii) Irrigation efficiency with surface irrigation systems may be as high as

(a) 50%
(b) 60%
(c) 70%
(d) 80%

(viii) The fraction of irrigation water that drains beyond the root zone is called

(a) Leaching requirement
(b) Leaching fraction
(c) Permissible leaching
(d) Steady-state leaching

(ix) Water depth during each irrigation in saline soils equals the sum of

(a) Leaching requirement and leaching fraction
(b) Leaching fraction and crop water requirement
(c) Leaching requirement and crop water requirement
(d) Leaching fraction and transpiration

(x) To increase crop WUE in sandy soils, irrigations should be

(a) Shallow and frequent
(b) Deep and frequent
(c) Shallow and less frequent
(d) Deep and less frequent

(xi) Soils are generally not suited to surface irrigation systems if their infiltration rates are

(a) >10 mm/h
(b) >50 mm/h
(c) >100 mm/h
(d) >200 mm/h

[**Key:** (i) b, (ii) a, (iii) d, (iv) c, (v) a, (vi) d, (vii) d, (viii) b, (ix) c, (x) a, (xi) c]

References

Corwin DL, Rhoades JD, Simunek JS (2012) Leaching requirement: steady state versus transient models. In: Wallender WW, Tanji KK (eds) Agricultural salinity assessment and management. ASCE Publ, pp 801–824

Fipps G (2003) Irrigation water quality standards and salinity management. Fact Sheet B-1667. Texas Cooperative Extension. The Texas A&M University System, College Station, TX

Gilley J, Watts D (1977) Possible energy savings in irrigation. J Irri Drainage Div Proc ASCE 103:445

Kan I, Rapaport-Rom M (2012) Regional blending of fresh and saline irrigation water: is it efficient? Water Resour Res 48(7):1–14

Koech R, Langat P (2018) Improving irrigation water use efficiency: a review of advances, challenges and opportunities in the Australian context. Water 10:1771–1788

Postel S (1999) Pillar of sand: can the irrigation miracle last? W.W. Norton and Co, New York, NY

Rhoades JD, Kandiah A, Mashali AM (1992) The use of saline waters for crop production. FAO irrigation and drainage paper 48. Food and Agriculture Organization of the United Nations, Rome

Sang H, Guo W, Gao Y, Jiao X, Pan X (2020) Effects of alternating fresh and saline water irrigation on soil salinity and chlorophyll fluorescence of summer maize. Water 12(11):3054

Seckler D, Molden D, Sakthivadivel R (2003) The concept of efficiency in water resources management and policy. In: Kijne W, Barker R, Molden D (eds) Water productivity in agriculture: limits and opportunities for improvement. CAB International, pp 37–50

Siebert S, Kummu M, Porkka M, Döll P, Ramankutty N, Scanlon BR (2015) A global data set of the extent of irrigated land from 1900 to 2005. Hydrol Earth Syst Sci 19:1521–1545

Sivanappan RK (1994) Prospects of micro-irrigation in India. Irrig Drainage Systems 8:49–58

Taylor SA, Ashcroft GL (1972) Physical edaphology. WH Freeman and Co., San Francisco, pp 434–435

Chapter 4
Drainage

Optimum plant growth and yield require a balance of air and water in the root zone. Excess water per se is not toxic to plants; otherwise, plants would not grow in hydroponics, but it harms plant growth by bringing undesirable changes in the root zone, such as O_2 stress, CO_2 toxicity, nutrition problems, and production of plant toxins. These soil conditions are unfavorable for healthy plant growth. Even short-term flooding for a day or so may kill plants and significantly reduce crop yields. Soil saturation also interferes with tillage and other cultural operations in the field because of operational problems of farm machinery. Drainage is a problem of irrigated lands in humid as well as in arid and semiarid regions of the world. Today, nearly 275 million hectares of agricultural lands are irrigated, which account for nearly 20% of the cultivated lands globally (WWAP, 2017). Several of such lands have developed drainage problems due to varied reasons. To make them productive, the excess water must be drained. This chapter discusses drainage problems, causes of excess water conditions in agricultural lands, different drainage systems, and impacts of drainage on soil and plant growth.

4.1 Drainage

Drainage may be defined as "the artificial removal of excess water from land." Majority of the agricultural lands are naturally drained through surface runoff and internal drainage. *Natural drainage* is a process by which water moves across, through, and out of the soil as a result of the force of gravity. Naturally drained water supports seeps, springs, stream baseflow, and aquifer recharge. Artesian water is also an example of natural flow of water by gravity in confined aquifers. Internal drainage moves excess water downwards beyond the root zone, and in many agricultural soils, it is sufficient to prevent waterlogging. The excess water may flow

P. K. Sharma, S. Kumar, *Soil Physical Environment and Plant Growth*,
https://doi.org/10.1007/978-3-031-28057-3_4

from crop fields to swamps or lakes and rivers or to the groundwater. In some lands, however, natural drainage is not enough to keep soil in aerated condition. Such lands require artificial removal of water from surface or subsurface to facilitate crop production.

4.1.1 *Objectives of Drainage*

There are three main objectives of drainage:

(i) To reduce soil submergence, i.e., removal of excess water from soil
(ii) To control salinity, i.e., removal of excess salts (and phytotoxins) from soil
(iii) To improve agricultural productivity of land

Besides, drainage also prolongs the length of cropping season, facilitates tillage operations, maintains soil structure, and makes new/additional land accessible for agriculture.

4.1.2 *Characteristics of Drainage Condition*

The drainage condition of soils can be easily judged from the following characteristics:

(i) *Water stagnation in the field*: Is water stagnation continuous or occasional?
(ii) *Soil water content*: Is soil saturated or near saturated? What is the duration of saturation or near saturation?
(iii) *Color of soil profile*: Is the soil brightly colored or grayish or yellowish in color? The soils with good drainage are brightly colored, usually reddish, brownish, or buff color due to different degrees of oxidation of iron; soils with poor drainage are mottled, gray, or yellowish in color.

4.2 Causes of Waterlogging/Soil Saturation

A part of the water applied as irrigation or received as rainfall enters into soil, and a part is stored at soil surface depending on the infiltrability of soil. Under prolonged irrigation or rainfall events, especially in low-permeable soils, pools of water start forming on the soil surface, and the process is called water ponding. The *water ponding* refers to "the accumulation of water on the soil surface." A part of the infiltrated water is stored in soil, and the rest undergoes deep percolation. If the percolation water reaches the saturated soil, the water table starts rising. Shallow water tables, capable of keeping root zone saturated/nearly saturated through capillarity, harm the plant growth and crop yields. Such a soil is called a waterlogged soil.

Waterlogging refers to "the accumulation of excess water in the root zone of the soil." Ponding of water and waterlogging may be temporary (for a few hours/days/weeks/months in a cropping season) or permanent (for the entire cropping season or more) and is undesirable in agricultural lands.

Waterlogging is generally a problem of low-permeable soils in low-lying areas experiencing heavy-rainfall events and having shallow water tables and poor drainage conditions. Over-irrigation may also result in waterlogging. Huge agricultural lands worldwide are suffering from drainage problems. Nearly 16% of agricultural lands in the United States, 10% of the agricultural lands of Russia, and sizeable irrigated lands of India, Pakistan, Bangladesh, and China are excess-water stressed (FAO 2015). The climate change has further aggravated the drainage problems. Waterlogging events have become more frequent, severe, and unpredictable due to climate change (IPCC 2014). Wet areas are expected to become wetter with prolonged waterlogging.

One or more of the following factors are responsible for waterlogging of agricultural lands:

- The physiography of land (topography, slope, shape, and drainage patterns) plays an important role in waterlogging by determining the rate of surface runoff. Low-lying areas such as valleys, depressions, and flat lowlands are more prone to waterlogging than the flatlands.
- Heavy and consistent rains especially in low-permeable soils and in soils with textural stratification (sandy top- and clayey subsoils) cause surface flooding. In permeable soils, such situations may raise the water table, resulting in waterlogging.
- The clayey soils having low permeability and easily crusting soils are prone to waterlogging.
- The geology determines the vulnerability of land to waterlogging. The shallow soils or soils having hardpans close to surface or soils having shallow water tables get easily waterlogged due to heavy-rainfall events.
- Interflows, subsoil flows (from upper regions to lower areas), and seepage from nearby water bodies like lakes, shallow aquifers, canals, and rivers promote waterlogging.
- Excessive irrigations and poor drainage systems often lead to surface flooding of agricultural lands.
- Amelioration of salt-affected soils in arid and semiarid regions requires flushing of salts from the root zone with irrigation water. The percolating water dissolves salts (and other toxins) from the soils and removes them through the subsurface drains. In the absence of proper drainage system, it raises the groundwater table. The flushing of salts from the root zone with the help of percolating water is called *leaching*.
- Shallow water tables through capillary action keep the root zone saturated or nearly saturated continuously or for a major part of the cropping season.

4.3 Characterization of Waterlogging in Relation to Plant Growth

Water tables shallower than 30 cm depth adversely affect the growth and yield of all arable crops (FDEU 1972). Using this concept, an index called *sum of excess waterlogging* in the top 30 cm soil layer (SEW_{30}) was developed by Sieben (1964) to judge the severity of waterlogging on crop yields. The index correlates well with crop yields (Cox and McFarlane 1995; Setter and Waters 2003).

The SEW_{30} refers to the "sum of excess water" that occurs each day in the top 30 cm soil layer. The SEW_{30} (cm-days) is computed as the sum of all daily values (cm) by which the water table is closer than 30 cm to the soil surface by using the following equation:

$$\mathrm{SEW}_{30} = \sum_{i=1}^{n} (T_i) \quad (4.1)$$

$$T_i = 30 - X_i \text{ for } X_i < 30\,\text{cm}$$

$$T_i = 0 \text{ for } X_i \geq 30\,\text{cm}$$

where X_i is the average depth of water table below the soil surface (cm) on day i ($i = 1, 2, ..., n$) and T_i is a statistic based on X_i values. A value of $X_i > 30$ cm means that the soil is not waterlogged.

For example, if the water table remains at 20 cm depth from the soil surface for 15 days, the ESW_{30} for the 15-day period ($i = 1$–15) is computed as

$$ESW_{30} = 15(\text{days}) \times (30 - 20\ \text{cm}) = 15 \times 10 = 150\ \text{cm} - \text{days}$$

Based on ESW_{30}, seven waterlogging classes (in terms of drainage conditions) have been identified (Table 4.1). The SEW_{30} indicates the potential threat of excess

Table 4.1 Waterlogging classes in terms of drainage condition using SEW_{30} index

Waterlogging class (in terms of drainage)	SEW_{30} index (cm-days in average growing season)
Well drained	<30
Moderately well drained	30–100
Moderately drained	100–250
Imperfectly drained	250–500
Moderately poorly drained	500–1200
Poorly drained	1200–2500
Very poorly drained	>2500

Source: Bulletin 4343 Soil Guide: a handbook for understanding and managing agricultural soils (Agriculture and Food, Department of Primary Industries and Regional Development, Government of Western Australia)

water conditions to plant growth. The actual impact of waterlogging on crop growth and yield, however, is influenced by several factors, viz. salinity, nutrient level, soil temperature, frequency of waterlogging, and crop growth stage.

4.4 Impact of Waterlogging/Soil Saturation on Soil Properties

Soil physical, chemical, electrochemical, and biological properties undergo significant changes due to waterlogging or soil saturation (Table 4.2).

4.4.1 Soil Structure

The immediate direct impact of waterlogging or soil saturation is the collapse of soil structure and creation of anaerobic conditions. Soil structure in sodic soils having high Na:Ca ratios is damaged due to dispersion of clay particles. In nondispersive waterlogged soils, the structure may collapse under their own weight. In saturated/nearly saturated soils, the use of farm machinery and livestock movement may damage soil structure. Soil structural damage leads to a series of changes in other soil physical properties including soil porosity, pore size distribution, and hydraulic conductivity.

Table 4.2 Impact of waterlogging on soil properties (Manik et al. 2019)

Soil properties	Impact of waterlogging
Physical changes	• Soil compaction • Increase in bulk density • Massive structural changes • O_2 depletion • CO_2 accumulation • Lowered diffusion coefficient for gases
Chemical changes	• Ion toxicity • Secondary metabolite toxicity
Electrochemical changes	• Increased specific conductance • Decreased soil redox potential (Eh) • Decreased soil pH
Biological changes	• Reduced mineralization • Reduced aerobic microbial activity • Reduced immobilization

4.4.2 Soil Air

The air in the pore space is completely replaced by water in waterlogged/saturated soils. The diffusion of atmospheric gases (including O_2) into soils is severely restricted due to their low diffusion rates in water. The diffusion rate of O_2 as well as CO_2 in water is about 1/10,000 times that in air. At 20 °C, diffusion coefficient of O_2 in air is 0.205 cm^2/s and in water is 0.180×10^{-4} cm^2/s, while the diffusion coefficient of CO_2 in air and water is, respectively, 0.161 cm^2/s and 0.177×10^{-4} cm^2/s. The solubility of CO_2 in water (0.878 g/L) is about 28 times higher than that of O_2 (0.031 g/L). The limited supply of O_2 in soil pores is depleted rapidly by root and microbial respiration, and soil reductants, and the concentration of CO_2 in soil solution is raised. Few hours of soil saturation, therefore, leads to O_2 stress, CO_2 toxicity, and decrease in soil pH and soil redox potential (Eh).

4.4.3 Soil Redox Potential and Nutrient Availability

In well-aerated soils, Eh is around 600 mV (i.e., >350 mV). Waterlogging causes reduction in Eh to values as low as −700 mV. At Eh ≤350 mV, O_2 is nearly absent. Lowering of soil Eh leads to denitrification (conversion of NO_3^- to N_2O or N_2) and reduction of Fe^{+3} to Fe^{+2}, Al^{+3} to Al^{+2}, and Mn^{+4} to Mn^{+2}. With further drop in Eh, i.e., at ≤ −220 mV, SO_4^{2-} is reduced to H_2S or S^{2-}, and CO_2 to CH_4. Eventually, H_2O is also reduced to H_2. The solution concentrations of reduced Fe, Al, and Mn cations in saturated soils are two to ten times higher than in well-aerated soils and are toxic to plants (Khabaz-Saberi et al. 2006). Waterlogged soils are deficient in plant nutrients such as N, S, and Zn. Microbial metabolism in anaerobic soils leads to production and accumulation of different organic acids (e.g., acetic acid, butyric acid, phenolic acid), which are toxic to plants. Anaerobic conditions are also responsible for the emission of *greenhouse gases* (GHGs), especially CH_4 and N_2O.

4.4.4 Organic Matter Decomposition and Soil Salinization

The rate of decomposition of organic matter is drastically reduced under anaerobic conditions. That is why *peat* and *muck* soils are generally found in wetlands. Waterlogging also results in soil salinization and development of soil acidity. *Soil salinization* refers to "the accumulation of soluble salts at the soil surface or near the soil surface to levels that are harmful to soil properties and plant growth."

4.5 Drainage Systems

The drainage problems are of two types: (a) surface drainage problems and (b) subsurface drainage problems.

Surface drainage problem is most common in areas with high rainfall intensities or in low-permeable soils or a combination of both. In addition to rainfall and irrigation, the runoff, seepage, or overflow from field or stream channels may also be the causes of surface drainage problems. The problem increases further in flatlands having uneven land surface without or with poor excess-water disposal channels. Surface drainage is also needed in irrigated lands.

Subsurface drainage problem usually occurs in low-permeable soils having shallow water tables, as well as in irrigated lands in temperate-humid and arid and semiarid regions. Subsurface drainage in the irrigated lands in arid zones is needed primarily to control soil salinity. Salinity is generally not a problem in humid regions.

Based on the nature of drainage problem, the drainage methods are also of two types: surface drainage and subsurface drainage. Further, the drainage may be temporary (i.e., occasional drainage) or permanent. Temporary drainage systems may be needed in lowland (e.g., rice) and upland (e.g., wheat) cropping systems.

4.5.1 Surface Drainage

Surface drainage refers to "the removal of excess water from land surface." In surface drainage, the excess water is diverted from the soil surface directly to streams under gravity, thereby reducing the amount of water entering into and through the soil. To facilitate the flow of water towards outlet (open channel or stream, etc.), the fields are provided with artificial slopes, and the process is called *land shaping* or *grading*. Field dykes, diversion ditches, or shallow field drains may also be constructed to divert excess water towards the outlet. Land grading and shallow field drains are practiced in flat or undulating lands to induce gravitational flow of water towards outlets. Shallow field drains are constructed randomly connecting the low-lying areas, the potential spots of water stagnation, or may be parallel placed having fixed or variable spacing. Land grading removes small ridges and furrows formed in the fields during tillage operations, makes soil surface smooth, and provides a gentle slope towards the outlet. It avoids surface flooding even for a short period of time. Ponded water can also be drained by pumping.

Land grading on small fields or in hilly terrains may be achieved manually, but in large farms, it is usually accomplished by operator-controlled machines or automatic laser levelers. Shallow or deep open ditches may be constructed with the help of pan scrapers, blade graders, or bulldozers.

4.5.2 Subsurface Drainage

The subsurface drainage refers to "the removal of excess water from the root zone as well as for lowering the water table." Subsurface drainage is achieved by constructing open deep ditches, digging mole drains, and installing drainpipes or well drainage. Excess water from the root zone flows into ditches, drains, or pipes under gravity and is safely removed.

Deep open channels are dug at suitable spacing in the field. The ditches should have stable side slopes to avoid their collapse and sufficient flow capacity. Regular maintenance is required in open ditches to avoid sedimentation and control the growth of unwanted vegetation in them. Growth of grasses, weeds, shrubs, or trees generally reduces the flow capacity of open drains.

Mole drains are used in clayey soils mostly in humid regions (New Zealand, Great Britain) to remove excess water from the root zone. The *mole drains* are unlined channels formed in subsoil by pulling a cylindrical foot called torpedo (generally 65–75 mm diameter), attached to a narrow vertical steel blade, through the soil. A plug (also called expander) is attached behind the torpedo to smear and compact the channel walls and maintain the channel size. The mole drains are generally formed at 400–600 mm soil depth to avoid damage by tractors or other farm machinery. The mole drains are spaced 2–5 m apart. Wider spacing lowers their performance. These drains usually function for only a few years. Their operational period largely depends on the natural stability of the soil. Mole drains are better than open channels for subsurface drainage.

Subsurface drains constructed with clay tiles (C-shaped tiles or clay pipes), concrete tiles, or plastic pipes (smooth or corrugated) are comparatively more effective and longer lasting. High-density polyethylene (HDPE) and polyvinylchloride (PVC) are the two commonly used plastics for fabricating plastic pipes. These pipes may be placed in trenches manually or with the help of machines. The clay and concrete tiles (or pipes) are usually 300 mm long and 50–100 mm in diameter. Water enters in these drains through the joints of tiles. The plastic pipes are long (up to 200 m), flexible, smooth or corrugated, and perforated. They are buried in soil with the help of machines.

To prevent the flow of soil particles into drains, sometimes filter material is placed around the drainpipes. In case of clay or concrete tiles, filter material is placed at their joints. The plastic pipes are wrapped in filter material before burying them in the soil. The filter material may be thin sheets of fiber glass or nylon or envelops of sand, gravel, or other granular porous material. The depth and spacing of subsurface drains depend on the texture of soil (Table 4.3) as well as type of crop (Acharya et al. 2019).

The open dug-wells or tube wells are also used for subsurface drainage, and this technique is called *vertical drainage*. The water in wells is normally pumped out or the wells may be connected to drains for discharge by gravity.

Comparison of open channel and buried pipe drainage systems

Table 4.3 Average depth and spacing of tile drains in various soil types

Soil type	Hydraulic conductivity (cm/d)	Spacing of drains (m)	Depth of drains (m)
Clay	0.15	10–20	1.0–1.5
Clay loam	0.15–0.50	15–25	1.0–1.5
Loam	0.5–2.0	20–35	1.0–1.5
Fine sandy loam	2.0–6.5	30–40	1.0–1.5
Sandy loam	6.5–12.5	30–70	1.0–1.5
Peat	12.5–25	30–100	1.0–2.0

Source: Ghildyal and Tripathi (1987)

- Open drain system is simple and less expensive, while buried pipe drain system is more complex and expensive due to the cost of drainpipes and installation cost due to the involvement of equipment and skilled manpower.
- Open drains consume cultivable land, which otherwise could be used by the crops, while there is no such loss of cultivable land in case of buried pipe drains.
- Open drains interfere with the movement of farm machinery, while no such problem exists with buried pipes.
- Maintenance cost is higher in case of open drains than in buried pipe drains. Open drains require frequent maintenance to control unwanted vegetation growing in the channels, but no such problem exists in buried pipes. There is also every chance of collapse of earthen walls of open channels, and the debris may block the water flow, reduce the water flow, or reduce the flow capacity of drains. No such problem exists with buried drains.

4.6 Drainage Equation

The drainage equation is used to design a subsurface drainage system by estimating the spacing of subsurface drains and the water flux into them. The drainage equation describes the relationship among depth and spacing of parallel subsurface drains, water table depth, and hydraulic conductivity of soil. The flow of groundwater to the drainage systems is analyzed based on the concept of steady-state water flow in a saturated soil. The steady-state condition assumes that the level of the water table remains constant and the discharge rate equals the rate of groundwater recharge. The steady-state condition is fulfilled if long-period drainage (say for a season or a year) is considered. Cross sections of two subsurface drainage systems are shown in Fig. 4.1.

Hooghoudt, a Dutch scientist, developed a drainage equation in 1940 for describing subsurface drainage. The equation is based on Darcy's law for saturated water flux under steady-state conditions. This equation, called the *Hooghoudt's drainage equation,* has been used for describing subsurface water flows and defining the spacing of subsurface drains. The equation is based on three assumptions:

(i) Subsurface flow is two-dimensional.

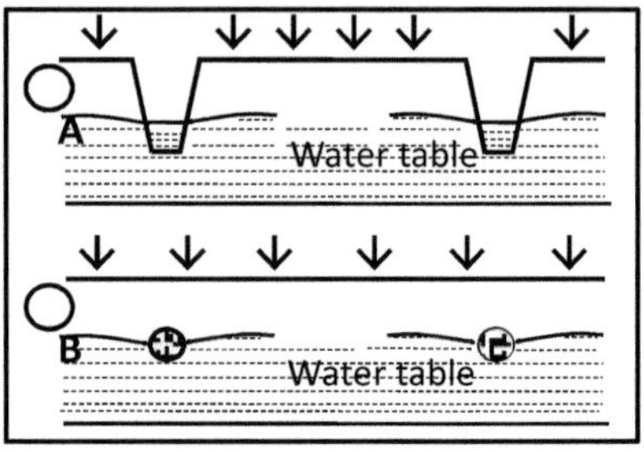

Fig. 4.1 Cross sections of open-field drains (A) and pipe drains (B), showing a curved water table under recharge (Source: Ritzema 1994)

(ii) The recharge of the water table is uniformly distributed.
(iii) Soils are homogeneous and isotropic. It ignores spatial variability in hydraulic conductivity of soils.

Wesseling (1973) modified the Hooghoudt's equation, which is now the most widely used equation. The modified equation is as follows:

$$qL = (8H_{\mathrm{m}} / L)(K_{\mathrm{b}}.D_{\mathrm{e}} + K_{\mathrm{a}}.H_{\mathrm{a}}) \tag{4.2}$$

where q is the steady-state drainage discharge rate recharge which is equal to the water percolating to the water table (m/d or m/h), L is the drain spacing (m), D_{w} is the height of water table midway between the center of two drains (m), K_{b} is the hydraulic conductivity of soil below drain level (m/d or m/h), K_{a} is the hydraulic conductivity of soil above drain level (m/d or m/h), D_{e} is the Hooghoudt's equivalent depth to the impermeable layer below drain level, and H_{a} is the average height of the water table above drain level (m).

The Hooghoudt's equivalent depth (D_{e}) is related to the drain spacing (L), depth (D) of impermeable layer below the drain (=$D_{\mathrm{i}} - D_{\mathrm{d}}$), and radius of drain ($R$) as follows:

$$\text{If } D < R: \quad D_{\mathrm{e}} = D$$

$$\text{If } R < D < L/4: \quad \mathrm{D_e} = D.L / \{(L - D^2) + 8D.L.\ln(D/R)\}$$

$$\text{If } D > L/4: \quad D_{\mathrm{e}} = L / 8\ln(L/R)$$

Various parameters used in Hooghoudt's drainage equation are shown in Fig. 4.2.

Hooghoudt's drainage equation is used to calculate the spacing between subsurface drains. Subsurface drain spacing varies by soil type and ranges from approximately 10 m for heavy-clay soils to as much as 50 m for highly permeable soils.

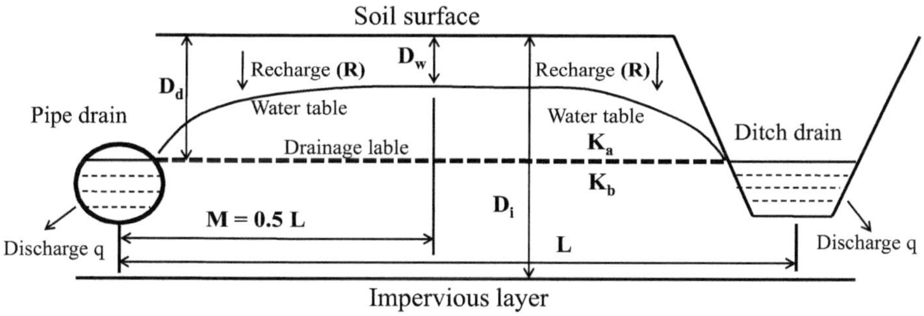

Fig. 4.2 Different parameters involved in Hooghoudt's drainage equation

4.7 Effect of Drainage on Soil Physical Properties

The direct effect of drainage is the lowering of water table, reduction in soil water content, and increase in gaseous exchange in the root zone. Drainage significantly improves soil structure and O_2 status of soils. Soil structural improvement is reflected in terms of increase in water-stable aggregates, macroporosity, and saturated hydraulic conductivity and decrease in surface runoff (Leyton and Yadav 1960; Hundal et al. 1976; Alakukku and Turtola 2010; Lokken et al. 2017).

The improvement in soil physical properties increases with the duration of drainage. Lokken et al. (2017) studied drainage impacts on soil physical properties in 42 wetlands for over 50 years and reported maximum changes in physical properties in soils drained for 20–34 years, depending on the soil type. After about 36 years of drainage, the changes in soil physical properties were stabilized in all soils. The impact of drainage on soil physical properties was influenced by other field practices, such as tillage, fertilizer application, and crop residue management. The permeability of soil below the drainpipes also influences changes in hydraulic properties of surface soils due to drainage (Frey et al. 2012).

4.8 Effect of Drainage on Plant Growth

Excess water is a significant abiotic stress to crop production in high-rainfall and/or poorly drained soils through its impact on soil properties and soil and plant processes. The conditions prevailing in waterlogged soils (O_2 stress, toxicity of organic and inorganic ions, deficiency of nutrients, etc.) do not support healthy plant growth and optimum crop yields. Several field crops including cereals, pulses, and oilseeds suffer yield losses by 20–50% or even more due to waterlogging (Manik et al. 2019). To make lands agriculturally productive, excess water has to be removed from the root zone by surface or subsurface drainage.

Plant growth and yield respond positively to drainage in waterlogged/poorly drained soils due to improvement in soil properties. Manik et al. (2019) have

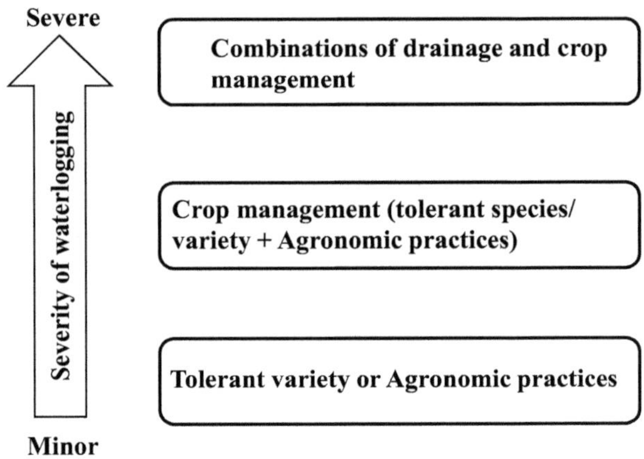

Fig. 4.3 Crop and soil management practices suggested for crop improvement on the basis of severity of waterlogging (Manik et al. 2019)

extensively reviewed the impact of soil and crop management practices, along with their advantages and disadvantages, on crop performance in waterlogged soils. They have recommended different management combinations depending on the severity of waterlogging (Fig. 4.3). Hundal et al. (1976) reported survival and performance of alfalfa and total hay production in the following order: surface and subsurface drained clay soil > subsurface drained > surface drained > undrained soil. Drainage has been found to bring changes in land-use pattern, farming systems, and cropping systems. For example, in England (eastern region), the 100% arable farming (mainly cereals) before drainage changed to 61% arable cereals, 37% arable potato and sugarcane, and 2% orchards (fruit plantation) after drainage. Similarly in Wales, farming system changed from 100% pasture before drainage to 44% pasture, 38% dairy farming, 16% mixed farming, and 2% arable cereals after drainage (FDEU 1972).

Water tables at depths >60 cm generally do not affect plant growth and yield. Water tables at 30–60 cm depth lower crop yields, but such yield losses can be compensated by good crop management practices including N applications, etc. (FDEU 1972). Supplemental N fertilizer can recover a part of yield loss caused by excess water because under O_2 stress conditions, nitrate can act as a secondary source of oxygen for the crop plants (Rajanna et al. 2018). The effects of soil drainage, use of fertilizer, and crop rotation on crop yields are additive (Bolton et al. 1982). Yield losses at water tables <30 cm depth are non-recoverable.

4.9 Question Bank

4.9.1 Short Questions:

(i) What is drainage? Why is it important in agriculture?
(ii) What are the objectives of drainage? Is it possible to judge the drainage situation in a field by simple field observations? If yes, how? If not, why?

(iii) What factors are responsible for waterlogging of crop fields?
(iv) What is SEW_{30}? How can you characterize waterlogging in relation to plant growth by using this index? Explain by giving a suitable example.
(v) What is the impact of waterlogging on soil properties, viz. soil structure, soil air, soil redox potential, nutrient availability, and organic matter decomposition?
(vi) Briefly explain different drainage systems commonly used to improve cultivated lands.
(vii) How does waterlogging affect plant growth?
(viii) Briefly describe *Hooghoudt's drainage equation* as modified by Wesseling (1973). What are the essential assumptions of this equation?
(ix) Compare and contrast the open channel and buried pipe drainage systems.

4.9.2 Briefly Explain Why?

(i) Drainage is essential to improve crop growth in waterlogged soils.
(ii) Color of soil profile is an important index of waterlogging situation of soil.
(iii) Prolonged irrigation or rainfall often results in waterlogging in low-permeable soils.
(iv) Waterlogging is generally a problem of low-permeable soils in low-lying areas.
(v) Immediate direct impact of waterlogging is the collapse of soil structure.
(vi) Few hours of soil saturation leads to O_2 stress and CO_2 toxicity in soils.
(vii) Waterlogging is responsible for emission of greenhouse gases.
(viii) Peat and muck soils are generally found in wetlands.

4.9.3 Fill in the Blanks:

(i) The artificial removal of excess water from land is known as ....
(ii) The process of removal of excess water from soil by force of gravity is known as ....
(iii) Bright color of soil profile is an indication of .... drainage condition.
(iv) The mottled, gray, or yellowish colored soil profile is an indication of .... drainage condition.
(v) The process of accumulation of water on the soil surface is known as ....
(vi) Soils having .... water permeability are less prone to surface water ponding.
(vii) Waterlogging is generally a problem of .... permeable soils.
(viii) SEW_{30} is an index of ....

(ix) SEW_{30} was developed by .... to judge the severity of waterlogging on crop yields.
(x) Soils having SEW_{30} index value of .... are classified as very poorly drained soils.
(xi) Solubility of O_2 in water is .... than that of CO_2.
(xii) The O_2 is nearly absent in soils at soil redox potential value of ....
(xiii) The soil redox potential of waterlogged soils may be as low as ....
(xiv) The spacing of tile drains is generally .... in sandy than in clay soils.
(xv) The depth of tile drains irrespective of soil texture generally varies between ...... and ..... meters.

[**Key:** i. drainage, ii. natural drainage, iii. good, iv. poor, v. water ponding, vi. high, vii. low, viii. waterlogging, ix. Sieben 1964, x. more than 2500, xi. less, xii. <350 mV, xiii. −700 mV, xiv. higher, xv. 1 and 1.5]

4.9.4 State Whether the Following Statements Are True (T) or False (F):

(i) Waterlogging is generally a problem of sandy soils in humid regions.
(ii) Artesian water is an example of natural drainage in confined aquifers.
(iii) One of the objectives of drainage is to control soil salinity.
(iv) Well-drained soils are brightly colored.
(v) Increasing O_2 concentrations in soils leads to mottling of soil profiles.
(vi) Excessive oxidation of Fe in soils gives yellowish color to soil profiles.
(vii) The geology determines the vulnerability of land to waterlogging.
(viii) Water table <60 cm depth adversely affects plant growth and yield.
(ix) ESW_{30} is an index of waterlogging in relation to plant growth.
(x) The higher the value of ESW_{30}, the better the drainage condition of soil.
(xi) Waterlogging increases soil redox potential.
(xii) Waterlogging increases the rate of organic matter decomposition.
(xiii) Peat and muck soils are generally found in low-lying waterlogged areas.
(xiv) Land grading is an important step in subsurface drainage.
(xv) Mole drains are commonly used for surface drainage of soils.
(xvi) The spacing of tile drains decreases, while the depth of drains increases with the fineness of soil texture.
(xvii) Depth of tile drains generally remains almost the same, i.e., 1–1.5 m, irrespective of soil texture.
(xviii) The *Hooghoudt's drainage equation* describes subsurface drainage.

[**Key:** i. F, ii. T, iii. T, iv. T, v. F, vi. F, vii. T, viii. T, ix. T, x. F, xi. F, xii. F, xiii. T, xiv. F, xv. F, xvi. F, xvii. T, xviii. T]

4.9.5 *Multiple-Choice Questions:*

(i) Waterlogging is generally a problem of

(a) High-permeable soils
(b) Low-permeable soils
(c) Stratified soils
(d) Saline soils

(ii) Majority of agricultural lands are

(a) Naturally drained
(b) Artificially drained
(c) Subject to waterlogging
(d) Subsurface drainage

(iii) The primary objective of drainage is to

(a) Improve O_2 concentration soil
(b) Reduce CO_2 concentration in soil
(c) Reduce soil salinity
(d) Remove excess water from soil

(iv) Poorly drained soils have

(a) Reddish colored profiles
(b) Bright colored profiles
(c) Mottled and grayish profiles
(d) Buff colored profiles

(v) Subsurface drainage is required if water table is shallower than

(a) 30 cm
(b) 60 cm
(c) 90 cm
(d) 120 cm

(vi) The value of SEW_{30} index for well-drained soils is

(a) 0–30 cm-days
(b) 30–100 cm-days
(c) 100–500 cm-days
(d) 500–1000 cm-days

(vii) The depth of tile drains for subsurface drainage is generally

(a) 1.0–1.5 m
(b) 1.5–2.0 m
(c) 2.0–2.5 m
(d) 2.5–3.0 m

(viii) The *Hooghoudt's equation* is used to describe

(a) Recharge rate of water table
(b) Hydraulic conductivity below the drain level
(c) Surface drainage
(d) Subsurface drainage

(ix) If water table remains at 15 cm depth from soil surface for 20 days, the ESW_{30} for the 20-day period will be equal to

(a) 100 cm-days
(b) 200 cm-days
(c) 300 cm-days
(d) 400 cm-days

(x) Water table generally does not adversely affect plant growth if the water table depth is

(a) ≥30 cm
(b) ≥60 cm
(c) ≥90 cm
(d) ≥120 cm

[**Key:** i. b, ii. a, iii. d, iv. c, v. a, vi. a, vii. a, viii. d, ix. c, x. b]

References

Acharya U, Chatterjee A, Lee A, Daigh M (2019) Effect of subsurface drainage spacing and depth on crop yield. Agron J 111(4):1675–1681

Alakukku L, Turtola E. Surface runoff and soil physical properties as affected by subsurface drainage improvement of a heavy clay soil. 9th Intl Drainage Symp, CIGR and CSBE/SCGAB Proceedings 2010, Canada

Bolton EF, Dirks VA, McDonnell MM (1982) The effect of drainage, rotation and fertilizer on corn yield, plant height, leaf nutrient composition and physical properties of Brookston clay soil in southwestern Ontario. Can J Soil Sci 62:297–309

Cox JW, McFarlane DJ (1995) The causes of waterlogging in shallow soils and their drainage in southwestern Australia. J Hydrol 167:175–194

FAO (2015) The State of Food and Agriculture – Soial protection and agricultuire: breaking the cycle of rural poverty. Food and Agri Org United Nations, Rome, p151

FDEU. Annual Rep; 1972. Cambridge, UK

Frey SK, Rudolf DL, Parkin GW (2012) Spatial and temporal influences on hydraulic properties in macroporous tile-drained soil. Soil Sci Soc Am J 76(2):350–360

Ghildyal BP, Tripathi RP (1987) Drainage. In: Ghildyal BP, Tripathi RP (eds) Soil physics. New Age Int Publ, pp 381–400

Hundal SS, Schwab GO, Taylor GS (1976) Drainage system effects on physical properties of a lakebed clay soil. Soil Sci Soc Am J 40(2):300–305

Intergovernmental Panel on Climate Change [IPCC] (2014) Climate change 2014: synthesis report. Contribution of working groups I, II and III to the fifth assessment report of the intergovernmental panel on climate change. IPCC, Geneva

Khabaz-Saberi H, Setter TL, Waters I (2006) Waterlogging induces high to toxic concentrations of iron, aluminum, and manganese in wheat varieties on acidic soil. J Pl Nutr 29(5):899–911

Leyton L, Yadav JSP (1960) Effect of drainage on certain physical properties of a heavy clay soil. Eur J Soil Sci 11(2):305–312

Lokken R, Zhang Z, Comeau LP, Bedard-Haughn A (2017) Effects of drainage duration on mineral wetland soils in a prairie pothole agroecosystem. Soil Till Res 168:187–197

Manik SMN, Pengilley G, Dean G, Field B, Shabala S, Zhou MX (2019) Soil and crop management practices to minimize the impact of waterlogging on crop productivity. Front Plant Sci 10:140

Rajanna GA, Dass A, Paramesha V (2018) Excess water stress: effects on crop and soil, and mitigation strategies. Popular Kheti 6(3):48–53

Ritzema HP (1994) Subsurface flow to drains. In: Ritzema HP (ed) Drainage principles and applications, Int inst land reclamation and improvement (ILRI), vol 16. ILRI Publ, Wageningen, The Netherlands, pp 263–304

Setter TL, Waters I (2003) Review of prospects for germplasm improvement for waterlogging tolerance in wheat, barley and oats. Plant Soil 253:1–34

Sieben WH. Relationship between drainage conditions and crop yield for young light clay soils in the Nordost polder. Van Zee tot Lod. 40. Tjeenk Willink V, Zwolle, The Netherlands, 1964 (As cited by Skaggs, 1978.)

Wesseling J (1973; Publ 16: 2–56.) Subsurface flow into drains. Drainage principles and applications vol II: theories of field drainage and watershed runoff. ILRI, Wageningen, The Netherlands

WWAP (United Nations World Water Assessment Programme). (2017) The United Nations World Water Development Report 2017. Wastewater: The Untapped Resource. Paris, UNESCO.

Chapter 5
Soil Structure and Plant Growth

Soil is a heterogeneous, dynamic, and biologically active porous medium. Soil particles, called the soil solids, rarely exist as independent entities in soils. They are linked to each other through certain forces and cementing materials forming a three-dimensional framework, called the soil structure. Soil's properties and processes are intimately linked to soil structure. Soil structure is a dynamic and composite soil property that significantly influences physical, chemical, and biological processes occurring in soils, such as transport and storage of liquids, gases, and heat; root growth; microbial population and activities; and decomposition and storage of soil organic matter (SOM). A good soil structure provides drainage and offers good aeration, water movement and storage, and biological activity, which are vital for a healthy plant growth. Plants in a soil with a poor structure may suffer from water-logging, oxygen stress, nutrient deficiencies, and phytotoxins of organic and inorganic origin. Being a dynamic property, soil structure keeps on changing in response to various internal and external forces, including moisture and temperature fluctuations, biological activity, and human interventions. Compaction of agricultural lands associated with the use of heavy machines, for example, has become a serious challenge to agricultural mechanization. The root zone must possess good soil tilth and granular structure for optimum growth and yield of crops. Soil structural units, i.e., soil aggregates, hold a huge fraction of soil organic matter (SOM). They may serve as a large sink of atmospheric carbon dioxide addressing the issue of climate change. In this chapter, we shall study the genesis of soil structure, its evaluation, and impact on plant growth and management.

P. K. Sharma, S. Kumar, *Soil Physical Environment and Plant Growth*,
https://doi.org/10.1007/978-3-031-28057-3_5

5.1 Important Terminology

5.1.1 Soil Aggregate

Soil aggregate may be defined as "a group of two or more primary soil particles joined together and stabilized more strongly than the surrounding soil mass." Soil aggregate is an umbrella term which includes peds, crumbs, fragments, or clods.

5.1.2 Ped

A *ped* refers to "a naturally formed soil aggregate that can be clearly seen in the field." A ped may have one or more of the following characteristics:

- A ped has a characteristic shape.
- A ped is separated from other soil aggregates by a specific pattern of cracks.
- A ped has distinct surfaces over at least one-half of the aggregate.
- A ped has coatings over at least one-half of its surface.

5.1.3 Crumb

A *crumb* refers to "a soft, porous, more or less rounded ped, 1–5 mm in diameter".

5.1.4 Fragment

The *fragments* refer to "the pieces of a ped broken along the natural surfaces of weakness (called the *cleavage planes*) and having length generally between 5 and 100 mm."

5.1.5 Soil Clod

A *soil clod* refers to "an undefined lump of soil mass formed artificially in the field by means of physical disturbances (digging, ploughing, etc.) and held together relatively more strongly than the surrounding soil." Simply described, a soil clod is an aggregate having a length more than 100 mm.

5.1.6 Concretion

A *concretion* refers to "a local concentration of compounds which irreversibly cement the soil particles together," for example iron concretions and lime concretions.

5.2 Soil Structure

Soil structure is the architecture of soil. It refers to the organization and arrangement of soil particles inclusive of the resultant complex pore system. The *soil structure* may be defined as "the arrangement of primary and secondary soil particles into a definite pattern." The term "structure" involves two concepts, viz. the "fabric" and the "bonding." The *soil fabric*, also called *soil matrix*, refers to the arrangement of soil particles and pore spaces. The soil matrix is defined as "the continuous spatial network of soil particles and pore spaces." The *bonding* refers to "the interparticle forces binding the particles together."

The primary soil particles (sand, silt, clay, and organic matter) do not exist independently under natural conditions. They are joined together by means of various cohesive, adhesive, and electrostatic forces and stabilized by *cementing agents* forming *soil aggregates*. The aggregates are dominated by clay particles. Silt and fine sand particles may also be part of aggregates. The aggregates are separated from one another by planes of weakness, called the *cleavage planes*.

5.3 Soil Aggregation

Soil aggregation refers to the process of formation of soil aggregates. Soil aggregation may be defined as "the binding together of soil particles into secondary units." Soil aggregates are the soil structural units. Soil aggregates are formed through the combined action of aggregation and fragmentation processes. Several attractive and disruptive forces act on the soil particles to cause greater cohesion among some particles, and groups of particles, than others. The groups of soil particles so formed are called the *soil aggregates*. The soil aggregates are particles of different shapes, sizes, stability, and distinctness.

Aggregate formation involves two processes: (i) *flocculation* of primary soil particles and (ii) *stabilization* (i.e., *cementation*) of soil floccules. Bradfield (1936) aptly stated that "*aggregation is flocculation plus …!*", and this "*plus*" refers to "*cementation*." The flocculation is responsible for the grouping together of primary soil particles, and cementation is responsible for the stabilization of these groups into aggregates.

5.3.1 Flocculation

The *flocculation* is a process of "coming together of soil particles and forming floccules due to attractive forces." A process opposite to flocculation is called dispersion. The *dispersion* refers to "the separation of soil particles due to repulsive forces." Flocculation and dispersion are principally controlled by the attractive and repulsive forces of the electric double layer, which surrounds colloidal particles. When the attractive forces are dominant, flocculation occurs, and when repulsive forces are dominant, dispersion occurs. Clay particles, being the most active fraction of soil, play a major role in the flocculation or dispersion process.

5.3.2 Cementation

The *cementation* is a process by which soil floccules are stabilized due to the binding action of certain materials. The materials which bind soil particles together into aggregates are called the *cementing agents*. Important cementing agents are organic matter (humus), sesquioxides of iron and aluminum, clay, calcium and magnesium carbonates, and silica. The role of different cementing agents in aggregation varies with soil type and climatic conditions. For example, in ultisols, the oxides of Fe and Al primarily cause the formation of the micro-aggregates while organic matter plays a primary role in stabilizing the macroaggregates (Peng et al. 2015). Major micro-aggregating agents in tropical soils are Fe and Al oxides. In organic matter-rich soils, the micro-aggregating effect of oxides of Fe and Al may be masked by organic matter in combination with Ca and Mg (Opara 2009). The possible bonding mechanisms in soil aggregation involving quartz, clay, organic polymers, and cations have been described by Harris et al. (1965).

As per the aggregate hierarchy model, primary particles are bond together to form secondary particles (the micro-aggregates), and small secondary particles merge to form bigger peds or aggregates (the macroaggregates) (Edwards and Bremner 1964; Tisdall and Oades 1982).

5.4 Micro- and Macroaggregates

Soil aggregates, based on size, have been broadly classified into two categories: (i) *micro-aggregates* (<0.25 mm) and (ii) *macroaggregates* (>0.25 mm). The lower limit of micro-aggregates is considered to be equivalent to silt fraction or silt size aggregates, i.e., 0.02 mm (as per the ISSS system) and 0.05 mm (as per the USDA system). Accordingly, the size range of micro-aggregates would be 0.02–0.25 mm (ISSS system) or 0.05–0.25 mm (USDA system). Some workers also differentiate between *small micro-aggregates* (0.02–0.05 mm) and *large micro-aggregates* (0.50–0.25 mm) as well as *small macroaggregates* (0.25–2.0 mm) and *large macro-aggregates*, also called *mega-aggregates* (>2.0 mm) (O"Brien and Jastrow 2013; Totsche et al. 2018; Semenov et al. 2020). Ma'rquez et al. (2004) proposed another size range for micro- and macroaggregates, viz. *large macroaggregates* (>2 mm), *small macroaggregates* (0.25–2 mm), *micro-aggregates* (0.053–0.25 mm), and *mineral fraction* (<0.053 mm).

Micro-aggregates are formed from a wide variety of mineral, organic, and biological soil components. Oriented clay domains, micro-aggregates <0.002 mm, and/or coarse clay and fine silt particles coalesce around a central core of hyphal fragments and bacterial cells or colonies. The SOM; oxides/hydroxides of Fe, Al, Mn, and Si; aluminosilicates; carbonates, etc. act as gluing and cementing agents to stabilize the micro-aggregates. Polysaccharides have been found to be specifically relevant in binding micro-aggregates <0.05 mm. Around 40–60% of the microbial biomass in soil has been found to be associated with micro-aggregates (0.002–0.02 mm). Breakdown of macroaggregates also leads to the formation of micro-aggregates (aggregate hierarchy concept). Micro-aggregates are very resistant to mechanical disruption. Totsche et al. (2018) have published an extensive review on the formation and properties of micro-aggregates.

Macroaggregates are formed due to the joining of multiple micro-aggregates by roots and hyphae, and various transient polymeric substances like polysaccharides and proteins. Micro-aggregates are, thus, the building blocks of macroaggregates. The breakdown of macroaggregates produces micro-aggregates.

5.5 Description of Soil Structure

Soil structure has a broad spectrum of spatial arrangement of soil particles. There are two extremes: on the one hand is the *single-grain structure* characterized by randomly arranged soil particles without much cohesion, and on the other hand is the *massive structure* without discernible internal features. Within these two extremes is the *granular/aggregated* structure (Hillel 1998).

Soil structure is usually described by the size, shape, stability, and distinctness of structural units, i.e., aggregates, peds, or crumbs. Zakharov (1927) and Nikiforoff (1941) developed a qualitative description of soil aggregates, peds, and clods based on their shape, size, and surface roughness. Emerson (1967) characterized the soil aggregates based on their coherence (i.e., stability) in distilled water. Soil Survey Staff (1951) developed a quantitative evaluation system of soil aggregates based on the concepts of Nikiforoff (1941). Currently, this is the most commonly used system of classification of soil structure for field descriptions.

According to Soil Survey Staff (1951), the soil structure is classified based on the following three parameters:

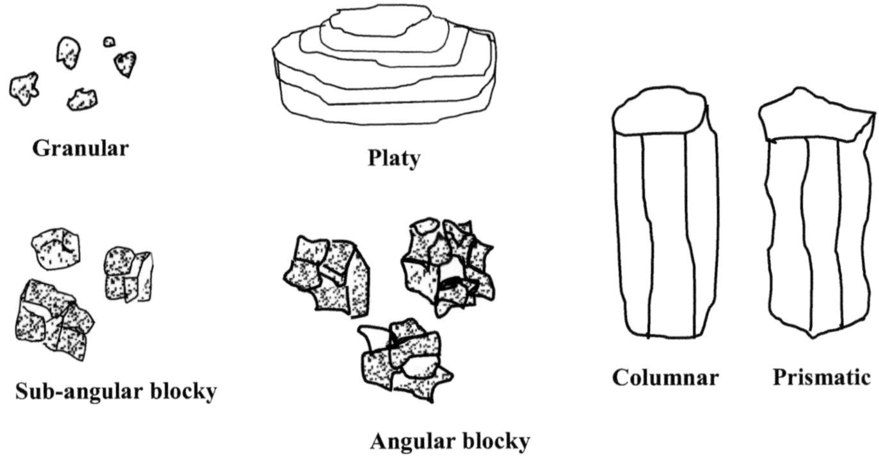

Fig. 5.1 Classification of soil aggregates as per their shape observed under field conditions

A. The *type* of soil structure (i.e., the shape and arrangement of peds)
B. The *class* of soil structure (i.e., the size of peds)
C. The *grade* of soil structure (i.e., the distinctness and durability of the peds)

A. The Type **of Soil Structure**

There are four *types* of structures based on the shape and arrangement of structural units (i.e., peds or aggregates) (Fig. 5.1).

A.1 Platy structure: Horizontal axes are longer than vertical axes with predominant horizontal cleavage planes. The structural units appear horizontally layered, thin, and flat, resembling wafers. Water flow is highly restricted. It is commonly found in forest soils, in part of the A horizon, and in claypan soils.

A.2 Prismatic or columnar structure: Vertical axes are longer than horizontal axes with predominant vertical cleavage planes. The structural units appear as vertically oriented pillars, often six sided, up to 15 cm in diameter. Vertical aggregates with top flats are called *prismatic*, and with tops rounded as *columnar*. Water flow is restricted leading to drainage problems. Such structures commonly occur in the B horizon of clay soils, particularly in semiarid regions.

A.3 Blocky structure: Vertical and horizontal axes are equally developed. The structural units appear as cube-like blocks up to 10 cm in size. The structure may be *angular blocky* (if corners and edges are sharp and distinct) or *subangular blocky* (if corners are rounded and edges are sharp). Water flow is restricted through these soils. These structures commonly occur in the upper part of B horizon, particularly where clay has accumulated.

A.4 Spherical structure: Also called granular and crumb structure. These are rounded aggregates, generally ≤2 cm in diameter. They are called *granules*. The porous granules are called *crumbs*. Spherical structure is often observed in A horizon. Water flow is easy through such soils.

Table 5.1 The class of soil aggregates based on the size (mm) of aggregates

Class	Platy	Prismatic	Columnar	Angular blocky	Subangular blocky	Granular	Crumb
1. Very fine or very thin	<1	<10	<10	<5	<5	<1	<1
2. Fine or thin	1–2	10–20	10–20	5–10	5–10	1–2	1–2
3. Medium	2–5	20–50	20–50	10–20	10–20	2–5	2–5
4. Coarse or thick	5–10	50–100	50–100	20–50	20–50	5–10	–
5. Very coarse or very thick	>10	>100	>100	>50	>50	>10	–

B. The class **of soil structure**

The *class* of soil structure is defined on the basis of average size of soil aggregates. Five distinct *classes* are recognized in relation to the *type* of soil structure:

- Very fine or very thin
- Fine or thin
- Medium
- Coarse or thick
- Very coarse or very thick

The average sizes of soil aggregates for any given *class* vary with the *type* of soil structure, as shown in Table 5.1.

C. The grade **of soil structure**

The *grade* of soil structure refers to the durability (or stability) of soil aggregates. There are four *grades* of soil structure:

C.1 Structureless grade: No observable aggregation or no definite orderly arrangement of natural lines of weakness (i.e., cleavage planes), such as:

- *Single-grain structure*: represents a structure where the individual soil particles show no tendency to cling together, such as pure sand
- *Massive structure*: represents a structure exhibiting tightly packed soil mass in large cohesive blocks, for example, structure of puddled (wet tilled) and dried rice soils

C.2 Weak grade: Poorly formed, nondurable aggregates, indistinct in undisturbed soil; upon disturbance, the soil mass breaks into a mixture of aggregates, broken aggregates, and unaggregated material.

C.3 Moderate grade: Well-formed, moderately durable aggregates, indistinct in undisturbed soil; upon disturbance, the soil mass breaks into a mixture of many distinct entire aggregates, some broken aggregates, and very little unaggregated material.

C.4 Strong grade: Well-formed, durable, distinct aggregates and quite evident in undisturbed soil; upon disturbance, the soil mass breaks into almost completely entire aggregates, very few broken aggregates, and no or very little unaggregated material; commonly known as granular structure.

5.6 Aggregate Stability

The a*ggregate stability* refers to "the ability of soil aggregates to resist disintegration due to water or wind or disruptive forces associated with tillage." *Wet aggregate stability* indicates the stability of soil aggregates against raindrop impact and water erosion. The size distribution of dry aggregates indicates resistance of aggregates to abrasive action of winds and wind erosion. The stability against tillage operations suggests the susceptibility of soil to mechanical disruption of aggregates or soil compaction.

Soil aggregation and aggregate stability are the result of complex interactions among various biological, chemical, and physical processes occurring in soil. Stability results from the rearrangement of soil particles, flocculation, and cementation of mineral particles with organic and inorganic substances. Several biotic (soil organic matter, microbial activity, earthworm activity, plant roots, etc.), abiotic (clay minerals, sesquioxides, exchangeable cations, etc.), and environmental factors (soil temperature, moisture, wetting and drying, freezing and thawing, etc.) affect aggregate stability (Chen et al. 1998; Sharma 2017). The soil aggregate stability is in fact the resultant of two opposing forces, i.e., the forces that bind the soil particles together and the forces that cause their disruption (disruptive stress). All those factors which impact soil aggregation also impact aggregate stability.

5.7 Evaluation of Aggregate Stability

The soil aggregates are evaluated in terms of size distribution, quantity, and stability of dry and wet aggregates. The process of evaluation of aggregates is known as *aggregate analysis*. Dry aggregates are analyzed by dry sieving, while wet aggregates are analyzed by different techniques such as wet sieving, raindrop impact technique, and sedimentation techniques. The choice of the method depends on the objective of the study. Dry sieving determines size distribution of dry aggregates, while wet sieving, raindrop impact, and sedimentation techniques determine size distribution of water-stable aggregates. Wet sieving is the most widely used method of water-stable aggregate analysis.

5.7.1 Dry Aggregation

Size distribution of dry aggregates is evaluated to determine the susceptibility of soil to wind erosion. Air-dry soil sample is sieved manually or with rotary sieve shakers (Chepil 1962) in a set of series of metallic sieves usually having 8.0, 4.0, 2.0, 1.0, 0.5, and 0.25 mm diameter holes. The soil aggregation may be expressed as *percent aggregation* or *mean weight diameter* (Sect. 5.8). The higher the percentage of larger aggregates (≥0.25 mm diameter), the lesser the susceptibility of soil to wind erosion.

5.7.2 Water-Stable Aggregates

Water stability of aggregates is determined to evaluate the stability of soil structure and the susceptibility of soil to water erosion.

5.7.2.1 Wet Sieving

Wet sieving is done under water in a set of metallic sieves, usually having 8.0, 4.0, 2.0, 1.0, 0.5, 0.25, and 0.10 mm diameter holes, using Yoder apparatus (Yoder 1936). Soil aggregates retained on different sieves after wet sieving are oven-dried and weighed. These aggregates are expressed in different ways (Sect. 5.8).

5.7.2.2 Raindrop Impact Analysis

The stability of aggregates is determined against simulated raindrop impact (McCalla 1944). Soil aggregates placed on a sieve are subject to falling drops of simulated rain of known diameter and from a fixed height. The kinetic energy used to break the aggregates is a measure of water stability of aggregates.

The kinetic energy of raindrops (E) used to shatter the aggregates is calculated as follows:

$$E = N \cdot mgh \tag{5.1}$$

where N is the number of raindrops required to shatter the aggregate, m the average mass of individual drop, g the acceleration due to gravity, and h the height of drop fall. If the density of water is considered as 1.0 g/cm^3, the mass of the drop (g) equals its volume (cm^3). The volume (V) of the drop having radius r (or diameter d) is computed by using the following equation of the volume of a sphere:

$$V = 4/3\pi r^3$$

$$\text{or } V = 4/3 \cdot \pi \left(d/2\right)^3 = 1/6\pi d^3 \tag{5.2}$$

The higher the energy required to break the aggregates, the higher the water stability of aggregates.

5.7.2.3 Sedimentation Techniques

Sedimentation techniques are used for the separation and estimation of finer fractions of water-stable aggregates. These techniques include *pipette method*, *hydrometer method*, and *elutriation method*. These are the standard techniques based on the relationship between size, density, and settling velocities of aggregates involving Stokes' law. Pipette and hydrometer methods are commonly used for particle size analysis. In the elutriation method, finer aggregate fractions are separated in a rising stream of water. The separation is based on the principle of balancing the settling velocity of aggregates and the upward flow of water stream. All aggregates with settling velocities more than the upward velocity of water in a particular vessel will settle down, and all others will be washed out of the vessel. Dry weights of different aggregate size fractions are determined.

5.8 Expression of Aggregate Data

Several indices have been developed to express aggregate stability data. Some of these indices are simple, viz. MWD (Van Bavel 1949), GMD (Mazurka 1950), instability coefficient (De Leenheer and De Boodt 1967), stability quotient (Sharma 1988), and WSA (%) (Kemper 1966; USDA 1998), while others are complex and elaborate, viz. aggregation index (van Steenbergen et al. 1991), normalized stability index (Six et al. 2000), and stable aggregate index (SAI) and stable macroaggregate index (SM_aI) (Ma'rquez et al. 2004). One may express aggregate stability using an index depending on the feasibility and objective of the study.

5.8.1 Water-Stable Aggregates (WSAs)

The WSA refers to "the percentage of water-stable aggregates >0.25 mm diameter." The index is calculated as follows (Kemper 1966; USDA 1998):

$$\text{WSA}(\%) = \{(\text{wt.of WSA} - \text{sand}) / (\text{wt.of dry soil sample} - \text{sand})\} \times 100 \qquad (5.3)$$

The water-stable aggregates (WSAs) are determined after wet sieving of soil sample. This index suggests the stability of aggregates against water erosion.

To determine the susceptibility of soil to wind erosion, dry aggregation is used for computing the index (Sect. 5.7.1). The index is called *percent aggregation* and refers to "the percentage of dry aggregates >0.25 mm diameter."

5.8.2 Mean Weight Diameter (MWD)

This index was introduced by van Bavel (1949). The MWD is computed as

$$\mathrm{MWD} = \Sigma x_i w_i \tag{5.4}$$

where *xi* is the mean diameter of any particular size range of aggregates and *wi* the weight of aggregates in that size range as a fraction of total dry weight of soil sample used in sieve analysis. MWD can be used for both dry and wet sieve analysis. It is the most commonly used index of aggregate stability.

5.8.3 Instability Index (ΔMWD)

Instability index is computed as the difference in MWD of dry and wet sieving analysis (De Leenheer and De Boodt 1967) and is represented as ΔMWD, i.e., change in MWD upon wet sieving:

$$\Delta\mathrm{MWD} = \mathrm{MWD}_{\mathrm{dry}} - \mathrm{MWD}_{\mathrm{wet}} \tag{5.5}$$

The smaller the value of this index, the better the soil structure.

5.8.4 Stability Quotient (SQ)

The *stability quotient* (SQ) refers to "the ratio of percent (dry) aggregation (>2 mm diameter) and change in mean weight diameter of aggregates upon wet sieving (ΔMWD)" (Sharma 1988). The index is computed as follows:

$$\mathrm{SQ} = \%\mathrm{SA} / \Delta\mathrm{MWD} \tag{5.6}$$

where % SA is the percent aggregation (dry) of soil. The higher the value of SQ, the better the soil structure. It is an improvement over instability index.

5.8.5 Geometric Mean Diameter (GMD)

Geometric mean diameter (GMD) is calculated by using the following equation (Mazurka 1950):

$$\mathrm{GMD} = \exp\left[\left(\Sigma w_i \log x_i\right) / \Sigma w_i\right] \tag{5.7}$$

where wi is the weight of the aggregates of average diameter xi and Σwi the total weight of the sample.

5.9 Soil Pore System

The soil pore system is the most important component of soil structure. Open spaces of varying shapes and sizes are found within the solid framework of soil particles (i.e., *soil matrix*). These open spaces are called the *soil pores* or *voids*. Soil pores are also created by plant roots, insects, earthworms, etc. when they move through soil. The pore geometry (i.e., size, shape, continuity, and tortuosity of pores) is very complex and is determined by the nature of particles forming soil aggregates and the mechanisms of aggregate formation. Water stability of aggregates determines the flexibility of soil pores.

Soil porosity refers to "the volume of voids within and between soil aggregates occupied by water and air." Soil porosity is a highly dynamic soil property, varying widely in space and time. Well-aggregated soil has higher total porosity than poorly aggregated soils. Soil porosity is an important index of soil structure. It has a significant effect on the transport and storage of water and air, transport and storage of plant nutrients, root growth, and microbial activities in soil.

5.10 Characterization of Soil Pore Space

The soil pore space may be characterized directly or indirectly by using different methods, such as pressure plate/membrane apparatus, mercury intrusion method, micromorphological method involving electro-optical image analysis, and X-ray imaging. The dependence of hydraulic properties on pore size, shape, and connectivity is widely used to make inferences about soil structure. Some of the characteristics of soil pore system are briefly described below:

5.10.1 Total Porosity

The *total porosity* (f) of soil refers to "the total pore volume (V_f) per unit volume of soil (V_t)." Mathematically,

$$f = V_f / V_t \tag{5.8}$$

Determination of volume of pores (V_f) in a given soil sample is rather difficult and involves special equipment. Hence, total porosity is generally calculated by using the following relationship:

$$f = 1 - \rho_b / \rho_s \tag{5.9}$$

where ρ_b is the bulk density and ρ_s the particle density of soil. Both bulk density and particle density are easy to be determined.

Total porosity is an indicative of water and air transmission properties of soil. A porous soil is good for proper drainage of root zone and optimum air exchange between soil and atmosphere.

5.10.2 Pore Size Distribution

Pore size distribution in soils is commonly determined by using the capillary rise equation:

$$d = 0.3 / h \tag{5.10}$$

where *d* is the diameter of the pore (cm) and h the matric suction (cm).

The volume of water retained between two suction values (h_1 and h_2) obtained from *desorption curve* prepared with the help of pressure plate or pressure membrane apparatus represents the volume of pores between the corresponding diameters (d_1 and d_2). The following example shall make the description clear:

Let a soil sample at matric suctions of 0.10 and 1.0 bar contains volumetric moisture content equal to 28% and 20%, respectively. The size of pores at the corresponding matric suctions would be

$$d_1 = 0.3 / 100\,\text{cm} = 0.003\,\text{cm} = 0.03\,\text{mm}$$

$$d_2 = 0.3 / 1000\,\text{cm} = 0.0003\,\text{cm} = 0.003\,\text{mm}$$

Thus, the volume of water retained by soil pores of sizes 0.003–0.03 mm dia would be (28–20 =) 8%. In other words, the given soil sample has 8% of the pores of sizes between 0.003 and 0.03 mm diameter.

In coarse-textured soils, pore size distribution can be measured with *pressure intrusion method*, in which mercury is forced into pores of oven-dried soil sample under pressure applied stepwise incrementally. The volume of mercury penetrating at each pressure step is measured, and the equivalent pore diameter is computed by capillary rise in Eq. 5.10 as stated above.

The pore size distribution has a great agricultural significance. It influences water and air permeability, plant available water capacity, root penetration, and growth and proliferation of soil microorganisms.

5.10.2.1 Classification of Soil Pores

Different pore size classification systems have been proposed by different investigators. The size limits sometimes appear to be logical and sometimes arbitrary. Few examples of pore sizes have been given below:

Example I

(i) Very fine pores: <2 μm (i.e., <0.002 mm)
(ii) Fine pores: 2–20 μm (i.e., 0.002–0.02 mm)
(iii) Medium pores: 20–200 μm (i.e., 0.02–0.20 mm)
(iv) Coarse pores: >200 μm (i.e., >0.20 mm)

Example II

(i) Cryptopores: <0.1 μm (i.e., <0.0001 mm)
(ii) Ultramicropores: 0.1–30 μm (i.e., 0.0001–0.03 mm)
(iii) Mesopores: 30–75 μm (i.e., 0.03–0.075 mm)

Cryptopores are so small that most of the organisms cannot enter into them. They help in preventing the decomposition of SOM. They hold water too tightly for plant use. *Ultramicropores* provide a good habitat for microorganisms. *Mesopores* are larger than macropores.

Example III
Pore size classification used by Russell (1973):

(i) Micropores: <0.01 mm diameter
(ii) Mesopores: 0.01–1.0 mm diameter
(iii) Macropores: >1.0 mm diameter

Mesopores and macropores are important for drainage, air exchange, and root growth.

Example IV
Greenland (1977) proposed pore size classification based on their significance to plant growth (Table 5.2).

The *storage pores* (0.50–50 μm) are the most important pore size range for plant growth. Water retained in pores ≤50 μm and ≤0.50 μm corresponds, respectively, to field capacity (FC) and permanent wilting point (PWP). Therefore, volume of pores in 0.50–50 μm size range corresponds to the plant available water capacity (FC-PWP). The *drainage pores* (>50–500 μm) and the *fissures* (>500 μm) are important for drainage, air exchange, and root growth. High percentage (70–80%) of *fissures* is indicative of poor soil structure for plant growth. High percentage of *storage* and *transmission/drainage* pores is indicative of a good soil structure.

Example V
Elliott and Coleman (1988) proposed four function-related categories of soil pores:

(i) Large macropores (root channels, earthworm holes, shrinkage cracks)
(ii) Inter-macroaggregate pores

Table 5.2 Functional classification of soil pores

Name of class of pores	Equivalent diameter (μm or 10^{-6} m)	Water potential (bar)	Function
Bonding space	<0.005	>−600	Important in terms of the forces holding domains and aggregates of primary particles together
Residual pores	0.005–0.5	−600 to −6	Chemical interactions at the molecular level
Storage pores	0.5–50	−6 to −0.06	Water storage for plants and micro-organisms
Transmission/ drainage pores	50–500	−0.06 to −0.006	Water and air fluxes, and root growth
Fissures	>500	<−0.006	Root penetration and drainage, especially in fine-textured soils

Adapted from Greenland (1977)

(iii) Inter-microaggregate pores
(iv) Intra-microaggregate pores

Example VI
The most commonly used system classifies soil pores into two categories: (i) micropores and (ii) macropores.

(i) *Micropores* (<80 μm, i.e., <0.08 mm): These are the intra-aggregate pores, i.e., the pores which occur within soil aggregates. The micropores are the major component of total soil porosity. They determine plant available water capacity of soil.
(ii) *Macropores* (>80 μm, i.e., >0.08 mm): These are the inter-aggregate pores, i.e., the pores which occur between soil aggregates. Macropores may result due to (i) shrinkage of soil during drying, (ii) soil cultivation, (iii) drying of old roots forming root channels, or (iv) earthworm activity forming burrows. The root channels and the earthworm burrows are called *biopores*. The biopores are more resistant to compaction due to the use of heavy agricultural machinery than the macropores formed by cultivation. Macropores are different from bulk soil physically, chemically, and biologically. They provide habitat for symbiotic as well as pathogenic microorganisms. Macropores are important for root growth, air exchange, and soil drainage.

5.10.3 Pore Geometry

The *pore geometry* refers to "the shape and size of pores, and their connectivity/ continuity in the system." The pore geometry is very complex and, thus, difficult to characterize.

- Soil pores are not a bundle of interconnecting cylindrical tubes of different diameters. They may be planar, completely irregular, cylindrical, or elongated in shape.
- The sizes are not uniform throughout the length of pores. A pore may widen and become narrow along its length. The narrowest point is important in terms of water and air movement, root growth, or microbial activity.
- All pores may not be interconnected through their entire length. Some pores may be blocked at places or may be poorly connected. These pores do not contribute towards water or air fluxes in soils and may create errors in the estimation of water fluxes using equations based on porosity values. Aggarwal and Sharma (1977) observed up to 5% blocked air pore volume in medium-textured soils using a simple method developed by them for the estimation of blocked air pore volume in soils.

5.10.4 Rigidity of Pores

The *rigidity* of pores refers to "the resistance offered by soil pores to deformation by growing roots." In simple words, it suggests how easily the soil pores can be deformed by the growing roots to make space for their growth. Compact soils have comparatively more rigid pore system. It is difficult for roots to grow and proliferate in compact soils.

5.11 Soil Structure and Plant Growth

Soil structure influences plant growth indirectly through its influence on four soil physical properties, viz. soil water, air, temperature, and penetration resistance. Soil pores, which are an important characteristic of soil structure, control these four parameters. Soil structure, and more so the soil pore system, is a highly dynamic physical soil property which keeps on changing with rainfall, irrigation, root growth, tillage, and other soil management practices. Because of the soil structural dynamism and complex interactions among four parameters, it is difficult to relate plant growth with soil structure precisely and reliably. Still many useful inferences with respect to soil structure and plant growth have been drawn by several studies. The most obvious effect of soil structure is on the root growth. The morphology and the water and nutrient absorption ability of roots are significantly affected by soil compaction. The root system, in turn, influences the shoot growth. Root and shoot systems have characteristic balance as dictated by the genetics of the plant species and are strongly influenced by soil structural characteristics. Any disturbance in root/shoot growth results in yield performance of the crop. The effect of soil structure on plant growth is briefly discussed below.

5.11.1 Root/Shoot Ratio

Soil structure affects plant growth mainly through its effect on root growth. The plant roots supply water, nutrients, and growth hormones to the shoots. In poorly structured soils, roots' ability to supply water and nutrients to shoots may be adversely affected. Roots may also send inhibitory signals to shoots in hard soils. Consequently, shoot growth is reduced.

Root/shoot ratio is generally genetically fixed for the particular cultivars. There is a general tendency within a given plant species to maintain a characteristic relationship between root and shoot dry weight. Root/shoot ratio is a complex parameter and is readily disturbed by several biotic and abiotic factors. Fertilizer applications, especially nitrogen, and irrigation can induce important changes in root/shoot ratio. Growth rates of roots and shoots during vegetation period continually keep adjusting to the soil and environmental conditions. Any reduction in shoot system (may be due to the removal of some parts of shoots) induces reduction in root growth to maintain the characteristic root/shoot ratio for that plant species. The higher root/shoot ratio indicates a robust root system and is in general a guarantee of more stable performance of plant species.

5.11.2 Root Growth and Root Growth Pressure

Root growth begins with the entry of root tips into soil pores. The meristem cells behind the root tip elongate and push the root tip forward. During elongation process, the roots experience (penetration) resistance offered by the soil matrix. The root elongation is controlled by the difference between the turgor pressure within the expanding cells and the penetration resistance of the soil. The resistance encountered by the elongating root tip in soil pores induces growth of root hairs on epidermal cells behind the meristem cells. The root hairs anchor the root tip and because of that most of the pressure of meristem cell elongation is transmitted to the root tip (Russell 1973). The pressure exerted by the growing roots enlarges/deforms soil pores to make space for the growing roots.

The roots must exert pressure greater than the mechanical strength of soil matrix to elongate. The pressure exerted by the plant roots to deform the soil pores is known as the *root growth pressure*. It is a combination of *axial pressure* and *radial pressure*. Root growth pressure results from the turgor pressure of meristem cells. Elongating/expanding root cells also have the ability to osmoregulate their (osmotic) pressure to counterbalance soil penetration resistance. The maximum *growth pressures* of *roots* from very different species are relatively similar (Clark and Barraclough 1999) and may vary between 0.9 and 1.3 MPa (Taylor and Ratliff 1969). Maximum axial growth pressures of roots are around 1.0 MPa (Taylor and Ratliff 1969; Hewitt and Dexter 1984), and maximum radial growth pressures are around 0.5 MPa (Gill and Bolt 1955). The differences in root growth pressure are largely associated with the differences in the root diameters (Dexter 1987). If the root growth pressure is more than the resistance offered by pore walls, the roots can enlarge the pores and continue growing. But if the resistance offered by pore walls is more than the pressure exerted by the growing roots, the roots fail to grow in soil and get deformed. For normal root growth, either the soil matrix must contain continuous pores between structural units which are large enough for roots to penetrate freely or the pores must offer little resistance against expansion by the growing roots.

The growing roots experience two types of pressures in soil: the *axial pressure* and the *radial pressure*. The axial pressure (i.e., the penetration resistance) is experienced on the root apex, and radial pressure is perpendicular to the root surface. To counterbalance these pressures, roots develop equivalent axial and radial pressures (collectively called the "root growth pressure"). Root growth is particularly sensitive to axial pressure, and the axial root growth pressure is very sensitive to the soil water potential. In one study, axial root growth pressure was observed to decrease from 0.65 to 0.35 MPa as the external water potential decreased from 0 to −0.4 MPa (Whalley et al. 1998). Axial pressure produces shorter and thicker roots, while radial pressure causes root bending. Many experts suggest, while others negate, the role of plant hormone, ethylene, in inducing modifications in the shape of impeded roots in compact soils (Passioura 1991). It may, however, not be incorrect to conjecture that both physical and biochemical factors are responsible for modifications in root morphology in compact soils. Root growth, in general, is impeded at bulk densities of 1.55 Mg/ m^3 in clay loams, 1.65 Mg/m^3 in silt loams, 1.80 Mg/m^3 in sandy loams, and 1.85 Mg/m^3 in loamy fine sands (Bowen 1981). Bulk density greater than 1.2 Mg/m^3 in clay soil, 1.6 Mg/m^3 in loam soil, and 1.8 Mg/m^3 in sandy loam adversely affects root growth of rice (Kar et al. 1976).

One may conclude, therefore, that roots growing in hard soils are morphologically different from the ones in friable soils. The roots are shorter, thicker, stubby, and clumped in hard soils. They are also much closely appressed to the soil particles. Roots are smooth and cylindrical in friable soils. The morphologically deformed roots are unable to satisfactorily absorb water and nutrients from soil. Many studies also suggest that plants grow more slowly in hard than in loose soils even if adequate supply of water and nutrients is maintained in the root zone (Passioura 1991). It is argued that roots under any kind of constraint (compact soil or water stress) send signal to the shoots in the form of a chemical (abscisic acid) along with water to reduce the expansion rate and stomatal conductance.

5.11.3 Pore Size and Root Growth

The pore space characteristics (pore size distribution, rigidity of pores, and continuity of root-sized pores) have a significant effect on root growth. Three interactions between a penetrating root and soil are commonly observed: (i) roots can penetrate into small pores by deforming soil to make a new root channel, (ii) roots elongate along an existing pore (or crack), or (iii) roots negotiate a crack at some angle of incidence before penetrating the soil.

The finest roots, i.e., root hairs, are generally 8–12 μm (0.0080–0.0012 mm) in diameter. Yamaguchi (2002) observed that in many field crops, the finer roots, i.e., root hairs (around 10 μm diameter), have higher frequency than the thicker roots. The percentage of root hair in wheat, maize, soybean, sugar beet, and potato was 60, 60, 30, 30, and 21, respectively. Around 88–97% of the roots ranged between 21 and 160 μm diameter. A large portion of these roots has diameters bigger than the soil pores, especially the *textural pores*, i.e., the pore space between the primary soil particles. Majority of roots, therefore, grow in macropores. Roots in macropores traverse deep and gain access to a larger reservoir of water and nutrients. The macropores may also enable the roots to traverse adequately even in otherwise compact soils, but in that case, the roots are clumped. The rate of water and nutrient extraction by clumped roots is slow. Such roots eventually may be unable to supply enough water or nutrients to the leaves, resulting in stunted plant growth.

A good root-soil contact is essential for optimum water and nutrient extraction. In friable soils, roots in macropores develop good contact with the solid and liquid phases for extraction of water and nutrients. Plants grow faster in soils having a large proportion of fine aggregates.

The significance of pore size distribution in plant growth varies with the soil texture. In sandy soils, large number of macro- and micropores is not good for plant growth as they allow excessive leaching of water and nutrients from the root zone. Fine-textured soils, on the other hand, have high microporosity, and the micropores hold water and nutrients too tightly to be easily available to plants. Loamy soils provide a healthy balance of micro- and macropores to support a healthy plant growth.

5.11.4 Factors Affecting Root Growth

The following characteristics of the root zone affect root growth:

- Physical properties: pore size distribution, soil penetration resistance, water, temperature, aeration, etc.
- Chemical properties: pH, nutrient concentrations, salinity, etc.
- Biological properties: root exudation, earthworms, mycorrhiza fungi, microbial activities, etc.
- Plant genetics and climatic conditions also have a significant impact on root growth.

5.12 Significance of Aggregate Stability in Agriculture

Aggregate stability is an important property of productive soils. It serves as an index of soil quality. The greater the proportion of water-stable aggregates, the better the soil quality. A healthy soil is one which has good aggregation, where stable pores extend from the soil surface to deep into the profile, and where pores allow water infiltration, root penetration, and air exchange to occur. Aggregate stability significantly affects soil erosion, movement of water, seedling emergence, and plant root growth. Stable aggregates generally provide a large range in pore space. Soil pores facilitate air exchange, water transport and storage, and chemical transport and storage and provide a habitat for soil flora and fauna.

A huge fraction of soil organic carbon is tied up in soils as an integral component of the structural units, i.e., soil aggregates. The carbon associated with soil aggregates is not easily exposed to decomposition. Soil structure, therefore, can serve as a reservoir of carbon and potential large sink for atmospheric CO_2. It has significance in processes leading to climate change.

The aggregate stability is a characteristic of a good soil structure. A *good soil structure* is one which has a stable aggregate fraction that can tolerate different disruptive forces, particularly water, and which provides continuity of pores in the soil matrix capable of supporting air and moisture exchange between the roots and soil environment. One of the main benefits of good soil structure is that it increases the duration of the cultivation window. It also minimizes tillage costs in terms of tractor hours, number of passes, and size of tractor and implements required.

5.13 Soil Tilth

Soil tilth refers to the physical condition of soil as related to its ease of tillage, fitness as a seedbed, seedling emergence, and root penetration. Hadas (1997) defined soil tilth as "the structural state of seedbed resulting from tillage and affecting crop response." The Soil Science Society of America (1996) defined soil tilth as "the physical condition of soil as related to its ease of tillage, fitness as seedbed, and its impedance to seedling emergence and root penetration."

Tilth is an important indicator of soil quality. It suggests the quality of soil structure in the seedbed or root zone. A good tilth must meet biological, agronomic, and engineering objectives of crop production. Tilth is difficult, if not impossible, to quantify. A good soil tilth in general is characterized by large pore space, rapid water intake, satisfactory water retention, good drainage, adequate aeration, least resistance to seedling emergence and root growth, and maximum resistance to soil erosion.

5.14 Maintenance of Stable Soil Structure

Soil structure (i.e., good soil tilth) can be sustained, improved, or destroyed by many physical (tillage, etc.), chemical (use of chemical fertilizers, soil conditioners, etc.), and biological (use of organics, etc.) practices and processes depending on the way the soil is managed. The practices that provide soil cover, protect or result in the accumulation of organic matter, maintain healthy plants, and avoid compaction improve soil structure and increase macropores. Management practices important for stable soil structure are briefly discussed below:

5.14.1 Practices that Improve Soil Structure

5.14.1.1 Tillage

Tillage practices significantly impact SOM and aggregate stability by bringing about three major changes in soil (Mehra et al. 2018):

(i) Changes in the soil climate that directly affect microbial activity and SOM
(ii) Changes in soil structure through physical disruption of aggregates and compaction
(iii) Changes in biodegradation rates when aboveground residues are incorporated into the soil and aggregate-bound carbon is exposed to the decomposing forces

For a given soil and tillage implement, the quality of tilth depends primarily on soil moisture content at the time of tillage. Tillage under wet conditions puddles the soil, which upon drying results in plastically deformed massive soil structure. These soils are hard to till, sometimes beyond the draft power of animals or small implements, and difficult to work into a suitable seedbed. When tilled, they end up into cloddy rough tilth. It is particularly true in fine-textured soils. Puddled rice soils are typical examples of this. Tillage under dry conditions causes minimal plastic deformation and structural damage but produces undesirable large fragments or fine dust depending on the intensity of tillage. The best soil moisture content for tillage is near the *lower plastic limit* when soil is in the *friable* state. Tillage at this moisture content produces granular seedbed of optimum tilth with minimum plastic deformation.

Historically, tillage was thought to create good soil structure for crop production by breaking up compact and hard soils. But recent studies have negated this thought. Studies have shown that, although initially tillage breaks up the soil and reduces compaction, in the long term, it actually increases soil compaction and is not good for soil structure (NRCS 2018). The conservation tillage (CT), including no-till (NT)/zero-till (ZT) system, improves aggregation and increases SOC levels, especially near the soil surface, due to minimum disturbance of soil aggregates and organic matter (Denef et al. 2004; Mehra et al. 2018). Soil enzymes are also increased under CT, which increases aggregate stability. Conventional tillage promotes loss of SOC through breakage of macroaggregates and exposure of SOC to microbial decomposition. The organic carbon bound in soil aggregates is resistant to microbial decomposition/oxidation. Organic carbon in the micro-aggregates is less susceptible to change than it is in the macroaggregates (Cambardella and Elliott 1993). Tillage operations also loosen the soil and increase aeration, which increases carbon mineralization. Reducing the frequency of disruptive tillage is also associated with increases in soil biological activity, an important factor of stable soil

Table 5.3 Effect of tillage systems on aggregation of four Indiana soils after 5 years (Mannering et al. 1975)

	Aggregation index	
Tillage system	0–5 cm depth	5–15 cm depth
Moldboard plow	0.35	0.47
Chisel plow	0.46	0.56
Till plow	0.47	0.56
No tillage	0.77	0.70

structure. The effect of different tillage systems on soil aggregation in four Indiana soils is shown in Table 5.3. Soil aggregation decreased as the tillage intensity increased. Highest aggregation was observed in the 0–5 cm soil layer under no-till system.

Tilling land at optimum soil moisture content when soil is in the friable range improves soil aggregation. Deep tillage, subsoiling, etc. loosen dense subsoils and other types of hardpans and improve soil structure.

5.14.1.2 Preservation of Soil Organic Matter

Preservation of soil organic matter (SOM) is crucial for improving soil aggregation and aggregate stability. Soil management practices that conserve SOM result in improved aggregate stability. Use of organic manures improves soil aggregation, while exclusive use of chemical fertilizers may destroy soil aggregation. Management of organic residues (crop residues, green manures, animal manures, etc.) stimulates microfaunal and floral activities in soil, improves soil SOM status, and favors aggregation and aggregate stability. The regular supplies of organic residues also minimize deleterious effects of intensive tillage on soil structure.

5.14.1.3 Agronomic Practices

There are several agronomic practices that help improve and sustain soil structure:

- Use of cover crops or adoption of crop rotations involving grasses, forages, or other crops having dense fibrous root system improves soil aggregation and promotes aggregate stability.
- Adoption of organic farming promotes and preserves soil aggregation.
- Maintenance of soil fertility by implementing efficient nutrient management systems (INM) helps sustain good soil structure.
- Adoption of environmentally acceptable drainage programs improves structure in waterlogged soils.
- Use of soil amendments like gypsum and soil conditioners improves and maintains aggregate stability.
- Proper grazing by animals is important for maintaining good soil structure, especially on the soil surface.

5.14.2 *Practices that Spoil Soil Structure*

The following agricultural practices lead to destruction of soil structure:

- Intensive soil disturbance and tillage operations may enhance the breakdown of soil organic matter, prevent accumulation of soil organic matter, and physically disrupt the existing soil aggregates.
- Cropping, grazing, or other production systems leave soil bare and expose it to the physical impact of raindrops or wind-blown soil particles.
- Removal of sources of organic matter and surface roughness by burning, harvesting, or otherwise removing crop residues.
- Using pesticides harmful to beneficial soil microorganisms.

5.15 Question Bank

5.15.1 *Short Questions:*

(i) Explain the following terms:

   (a) Soil aggregate
   (b) Ped
   (c) Crumb
   (d) Fragment
   (e) Soil clod
   (f) Concretion

(ii) Define soil structure. What is its significance in agriculture?

(iii) Define soil aggregation. Briefly explain the genesis of soil aggregates.

(iv) Differentiate between micro- and macroaggregates. What is their significance in crop production?

(v) How is the soil structure characterized according to the Soil Survey Staff (1951) system?

(vi) What do you understand by aggregate stability? Differentiate between dry and wet aggregate stability.

(vii) What are water-stable aggregates? Briefly describe the method of their determination by wet sieving and raindrop impact.

(viii) Briefly describe the following:

   (a) Percent water-stable aggregates (% WSAs)
   (b) Mean weight diameter (MWD)
   (c) Instability index (ΔMWD)
   (d) Stability quotient (SQ)
   (e) Geometric mean diameter (GMD)

(ix) What do you understand by pore size distribution in soil? Explain pore size classification as proposed by Greenland (1977).
(x) Differentiate between pore geometry and pore rigidity. What is their significance in plant growth?
(xi) What is root growth pressure? How is it related to root growth?
(xii) What is soil tilth? What are important characteristics of a good soil tilth?
(xiii) Briefly describe the impact of tillage on soil structure.
(xiv) What agronomic practices

(a) Preserve or improve soil structure
(b) Spoil soil structure

5.15.2 *Briefly Explain Why?*

(i) Soil particles rarely exist as independent entities in soil matrix.
(ii) Soil aggregation is "flocculation plus ...."
(iii) Dry aggregation is important to determine the susceptibility of soil to wind erosion.
(iv) Soil structure is a highly dynamic soil physical property.
(v) Soil pore system is an important characteristic of soil structure.
(vi) The "storage pores" (Greenland 1977) are the most important pore size range for plant growth.
(vii) Pore geometry is very complex and difficult to characterize.
(viii) Rigid pores are not good for plant growth.
(ix) Root growth pressure is an important characteristic of a plant in relation to root growth.
(x) Conventional tillage adversely affects soil aggregates.
(xi) Conservation tillage helps in preserving/improving soil structure.
(xii) Animal grazing has a significant effect on soil structure.

5.15.3 *Fill in the Blanks:*

(i) A .... refers to a group of two or more primary soil particles joined together and stabilized more strongly than the surrounding soil mass.
(ii) A naturally formed soil aggregate that can be clearly seen in the field is called ....
(iii) The pieces of a ped broken along the cleavage planes and having a length generally between 5 and 100 mm are known as ....
(iv) A .... refers to a local concentration of compounds, which irreversibly cement the soil particles together.

(v) The continuous spatial network of soil particles and pore spaces is known as ....
(vi) The aggregates are separated from one another by planes of weakness called the ....
(vii) The process of formation of soil aggregates is known as ....
(viii) The process responsible for grouping of primary soil particles is known as ....
(ix) The process opposite to flocculation, which causes separation of primary soil particles, is called ....
(x) The materials which bind soil particles together into aggregates are called the ....
(xi) Soil aggregates having size <0.25 mm diameter are called ....
(xii) Soil aggregates having size >0.25 mm diameter are called ....
(xiii) The .... structure is characterized by randomly arranged soil particles without much cohesion.
(xiv) A soil structure which does not have any discernible internal features is known as .... structure.
(xv) The .... of soil structure refers to the shape and arrangement of structural units.
(xvi) The .... of soil structure refers to the size of aggregates.
(xvii) The .... of soil structure refers to the durability of aggregates.
(xviii) The ability of soil aggregates to resist disintegration due to water or wind or disruptive forces associated with tillage is known as ....
(xix) The percent aggregation (dry) refers to the percentage of aggregates having a size of .... diameter.
(xx) The water storage pores (Greenland 1977) have a size range of ....
(xxi) The drainage pores (Greenland 1977) which are important for drainage of soils have a size range of ....
(xxii) The pressure exerted by the growing roots to counter the penetration resistance of soil is known as ....
(xxiii) The root growth pressure must be higher than the .... for the roots to grow in soil.
(xxiv) The root hairs are generally of a size range of ....
(xxv) The physical condition of soil as related to its ease of tillage, fitness as a seedbed, seedling emergence, and root penetration is known as ....

[**Key:** i. soil aggregate, ii. ped, iii. fragments, iv. concretion, v. soil matrix *or* soil fabric, vi. cleavage planes, vii. soil aggregation, viii. flocculation, ix. dispersion, x. cementing agents, xi. micro-aggregates, xii. macroaggregates, xiii. single grain, xiv. massive, xv. type, xvi. class, xvii. grade, xviii. aggregate stability, xix. >0.25 mm, xx. 0.5–50 μm, xxi. 50–500 μm, xxii. root growth pressure, xxiii. soil strength *or* soil penetration resistance, xxiv. 8–12 μm, xxv. soil tilth]

5.15.4 State Whether the Following Statements Are True (T) or False (F):

(i) The soil structure refers to the arrangement of primary soil particles into a definite pattern.
(ii) The pieces of a ped broken along the cleavage planes and having a length generally between 5 and 100 mm are known as crumbs.
(iii) The process of formation of soil aggregates is called soil aggregation.
(iv) Soil aggregation involves two processes, viz. flocculation and cementation.
(v) Soil aggregation is the resultant of two opposing forces, i.e. attractive and repulsive forces.
(vi) Organic matter plays a primary role in the aggregation of tropical soils.
(vii) The lower limit of micro-aggregates is considered to be equivalent to silt fraction or silt size aggregates.
(viii) Around 40–60% of the microbial biomass in soil has been found to be associated with macroaggregates.
(ix) Single-grain structure is most desired in cultivated soils.
(x) Massive structure results from tillage of wet soils.
(xi) The *type* of soil structure refers to the shape of aggregates.
(xii) The *class* of soil structure refers to the size of aggregates.
(xiii) The *grade* of soil structure refers to the durability of aggregates.
(xiv) Blocky structure is highly permeable and desirable in cultivated soils.
(xv) The granular or crumb structure is good for plant growth.
(xvi) Wet sieving determines the water stability of aggregates.
(xvii) Rotary sieve shaker is a commonly used apparatus for determining the water stability of aggregates.
(xviii) Yoder apparatus is used to determine the percent aggregation (dry) in soils.
(xix) Raindrop impact analysis determines the water stability of aggregates.
(xx) The water retained in storage pores is available to plants and soil microorganisms.
(xxi) The variation in root growth pressure of different plant species is not very large.
(xxii) The root growth pressure has two components, viz. axial pressure and radial pressure.
(xxiii) Majority of plant roots grow in micro- and mesopores.
(xxiv) A huge fraction of soil organic carbon is tied up in soils as an integral component of the structural units.
(xxv) Soil tilth is difficult, if not impossible, to quantify.

[**Key:** i. F, ii. F, iii. T, iv. T, v. T, vi. F, vii. T, viii. F, ix. F, x. T, xi. T, xii. T, xiii. T, xiv. F, xv. T, xvi. T, xvii. F, xviii. F, xix. T, xx. T, xxi. T, xxii. T, xxiii. F, xxiv. T, xxv. T]

5.15.5 *Multiple-Choice Questions:*

(i) Soil structure is the arrangement of the following particles into a definite pattern:

(a) Sand, silt, and clay
(b) Sand, silt, clay, and organic particles
(c) Sand, silt, clay, organic particles, and their combinations forming secondary particles
(d) Sand, silt, clay, and cations

(ii) Soil fabric refers to the arrangement of

(a) Soil particles
(b) Soil pores
(c) Soil particles and soil pores
(d) Soil separates

(iii) Cementing agents are the materials which are responsible for the

(a) Flocculation of soil particles
(b) Stabilization of soil floccules
(c) Flocculation and stabilization of soil aggregates
(d) Soil dispersion

(iv) Aggregation in tropical soils is primarily caused by

(a) Organic matter
(b) Ca and Mg ions
(c) Organic matter, Ca and Mg ions
(d) Oxides of Fe and Al

(v) Micro-aggregates are the aggregates which are smaller than

(a) 0.025 mm diameter
(b) 0.050 mm diameter
(c) 0.25 mm diameter
(d) 0.50 mm diameter

(vi) The *class* of soil structure according to the Soil Survey Staff (1951) classification refers to

(a) The shape of peds
(b) The size of peds
(c) The durability of peds
(d) The shape and size of peds

(vii) The *grade* of soil structure according to the Soil Survey Staff (1951) classification refers to

(a) The shape of peds
(b) The size of peds

(c) The durability of peds
(d) The shape and size of peds

(viii) Mean weight diameter (MWD) index of soil aggregation was developed by

(a) Chepil (1962)
(b) Mazurka (1950)
(c) van Bavel (1949)
(d) Stokes (1851)

(ix) Mean weight diameter (MWD) index of soil aggregation was developed by

(a) Chepil (1962)
(b) Mazurka (1950)
(c) van Bavel (1949)
(d) Stokes (1851)

(x) The Yoder apparatus is used for

(a) Aggregate analysis
(b) Wet sieving of aggregates
(c) Dry sieving of aggregates
(d) Determining the slaking of aggregates in water

(xi) According to Greenland (1977), the water present in the following soil pores is readily available to plants:

(a) Macropores
(b) Transmission pores
(c) Residual pores
(d) Storage pores

(xii) Pore geometry refers to the

(a) Shape of pores
(b) Size of pores
(c) Shape and size of pores
(d) Shape, size, and continuity of pores

(xiii) Root growth pressure refers to the

(a) Pressure exerted by the growing roots in soil pores
(b) Pressure exerted by the soil pores on the growing roots
(c) Pressure developed in xylem tissues of growing roots
(d) Pressure developed in the epidermal cells of growing roots

(xiv) The thickness of root hairs generally varies between

(a) 8 and 12 μm diameter
(b) 12 and 16 μm diameter
(c) 16 and 20 μm diameter
(d) 20 and 24 μm diameter

(xv) Majority of roots grow in

(a) Cryptopores
(b) Mesopores
(c) Micropores
(d) Macropores

[**Key:** i. c, ii. c, iii. b, iv. d, v. c, vi. b, vii. c, viii. c, ix. b, x. b, xi. d, xii. d, xiii. a, xiv. a, xv. d]

References

Aggarwal GC, Sharma PK (1977) Measurement of blocked air-pore volume in soils (porous media). Proc Indian Natl Sci Acad 43A(2):105–109

Bowen HD (1981) Alleviating mechanical impedance. In: Arkin GF, Taylor HM (eds) Modifying the root environment to reduce crop stress. ASAE, St. Joseph, MI, pp 21–57

Bradfield R (1936) The value and limitations of calcium in soil structure. Am Soc Assoc Bull XVII:31–32

Cambardella CA, Elliott ET (1993) Carbon and nitrogen distribution in aggregates from cultivated and native grassland soils. Soil Sci Soc Am J 57:1071–1076

Chepil WS (1962) A compact rotary sieve and the importance of dry sieving on physical soil analysis. Soil Sci Soc Am Proc 76:4–6

Chen Z, Pawluk S, Juma NG (1998) Impact of variations in granular structures on carbon sequestration in two Alberta Mollisols. In: R Lal et al. (ed) Soil Processes and the Carbon Cycle. Adv Soil Sci CRC Press, Boca Raton, Fl, pp 225–243

Clark LJ, Barraclough PB (1999) Do dicotyledons generate greater maximum axial root growth pressures than monocotyledons? J Exp Bot 50:1263–1266

De Leenheer L, De Boodt M (1967) Aggregate stability determination by the change in mean weight diameter. In West-European methods for soil structure determination, West European working group on soil structure of the ISSC. Ghent, VI, pp 28

Denef K, Six J, Merckx R, Paustian K (2004) Carbon sequestration in microaggregates of no-tillage soils with different clay mineralogy. Soil Sci Soc Am J 68(6):301–310

Dexter AR (1987) Mechanics of root growth. Plant Soil 98:303–312

Edwards AP, Bremner JM (1964) Use of sonic vibration for separation of soil particles. Can J Soil Sci 44:366–366

Elliott ET, Coleman DC (1988) Let the soil work for us. Ecol Bull 39:23–32

Emerson WW (1967) A classification of soil aggregates based on their coherence in water. Aust J Soil Res 5:47–57

Gill WR, Bolt GH (1955) Pfeffer's studies of the root growth pressures exerted by plants. Agron J 47:166–168

Greenland DJ (1977) Soil damage by intensive arable cultivation: temporary or permanent? Philos Trans R Soc Lond B 281:193–208

Hadas A (1997) Soil tilth—the desired soil structural state obtained through proper soil fragmentation and reorientation processes. Soil Till Res 43:7–40

Harris RF, Chester G, Allen ON (1965) Dynamics of soil aggregation. Adv Agron 18:107–160

Hewitt JS, Dexter AR (1984) The behavior of roots encountering cracks in soil. 2. Development of a predictive model. Plant Soil 79:11–28

Hillel D (1998) Environmental soil physics. Academic Press, San Diego, CA

Kar S, Varade SB, Subramanyan TK, Ghildyal BP (1976) Soil physical conditions affecting rice root growth: bulk density and submerged soil temperature regime effects. Agron J 68(1):23–26

Kemper WD (1966) Aggregate stability of soils from western United States and Canada. USDA-ARS. Technol. Bull. No. 1355. U.S. Gov. Print. Office, Washington, DC.

Mannering JV, Griflith DR, Richey CB (1975) Tillage for moisture conservation. Am Soc Agr Engrs Pap no. 75-2523. St. Joseph, MI
Ma'rquez CO, Garcia VJ, Cambardella CA, Schultz RC, Isenhart TM (2004) Aggregate-size stability distribution and soil stability. Soil Sci Soc Am J 68:725–735
Mazurka AP (1950) Effect of gaseous phase on water-stable synthetic aggregates. Soil Sci 69:135–148
McCalla TM (1944) Water-drop method of determining stability of soil structure. Soil Sci 58:117–121
Mehra P, Baker J, Sojka RE, Bolan N, Desbiolles J, Kirkham MB, Ross C, Gupta R (2018) A review of tillage practices and their potential to impact the soil carbon dynamics. Adv Agron 150:185–230
Nikiforoff CC (1941) Morphological classification of soil structure. Soil Sci 52:193–211
NRCS (2018) Soil health. https://www.nrcs.usda.gov/wps/portal/nrcs/main/soils/health/
O'Brien SL, Jastrow JD (2013) Physical and chemical protection in hierarchical soil aggregates regulates soil carbon and nitrogen recovery in restored perennial grasslands. Soil Biol Biochem 61:1–13
Opara C (2009) Soil micro-aggregates stability under different land use types in southeastern Nigeria. Catena 79(2):103–112
Passioura JB (1991) Soil structure and plant growth. Aust J Soil Res 29:717–728
Peng X, Yan X, Zhou H, Zhang YZ, Sun H (2015) Assessing the contributions of sesquioxides and soil organic matter to aggregation in an Ultisol under long-term fertilization. Soil Tillage Res 146:89–98
Russell EW (1973) Soil conditions and plant growth, 10th edn. William Cloves and Sons, London: Longman
Semenov VM, Lebedeva TN, Pautova NB, Khromychkina DP, Kovalev IV, Kovaleva NO (2020) Relationships between the size of aggregates, particulate organic matter content, and decomposition of plant residues in soil. Eurasian Soil Sci 53:454–466
Sharma PK (1988) Interaction of $FeSO_4$ with polyacrylamide and urea formaldehyde in aggregating light textured soils. Soil Sci 146:185–191
Sharma PK (2017) Introduction to soil physics. Westville Publishing House, New Delhi
Six J, Elliott ET, Paustian K (2000) Soil structure and soil organic matter: II. A normalized stability index and the effect of mineralogy. Soil Sci Soc Am J 64:1042–1049
Soil Survey Staff (1951) Soil Survey Manual. USDA Handbook No. 18
Soil Science Society of America (1996) Glossary of soil science terms. SSSA, Madison, WI
Stokes GG (1851) On the effect of the internal friction of fluids on the motion of pendulums. Trans Cambridge Phil Soc 9:8–106
Taylor HM, Ratliff LF (1969) Root growth rates of cotton and peanuts as a function of soil strength and soil water content. Soil Sci 108:113–119
Tisdall JM, Oades JM (1982) Organic-matter and water-stable aggregates in soils. J Soil Sci 33:141–163
Totsche KU, Amelung W, Gerzabek MH, Guggenberger G, Klumpp E, Knief C, Lehndorff E, Mikutta R, Peth S, Prechtel A, Ray N, Kögel-Knabner I (2018) Microaggregates in soils. J Plant Nutr Soil Sci 181(1):104–136
USDA (1998) Soil quality test kit guide. USDA-ARS, Washington, DC
Van Bavel CHM (1949) Mean weight diameter of soil aggregates as a statistical index of aggregation. Soil Sci Soc Am Proc 14:20–23
van Steenbergen M, Cambardella CA, Elliott ET, Merckx R (1991) Two simple indices for distributions of soil components among size classes. Agric Ecosyst Environ 34:335–340
Whalley WR, Bengough AG, Dexter AR (1998) Water stress induced by PEG decreases the maximum growth pressure of the roots of pea seedlings. J Exp Bot 49:1689–1694
Yamaguchi J (2002) Measurement of root diameter in field-grown crops under a microscope without washing. Soil Sci Plant Nutr 48(4):625–629
Yoder RE (1936) A direct method of aggregate analysis and a study of the physical nature of erosion losses. J Am Soc Agron 28:337–351
Zakharov SA (1927) Russian Pedology Investigations II. Acad. Sci. USSR

Chapter 6
Soil Air and Plant Growth

Soil air plays a vital role in modifying physical, chemical, and biological environment in soils, which significantly influences plant growth and yield. For healthy plant growth and microbial activities, air and water must be in proper balance in soil pores. The O_2 is essential for the survival of plants, while CO_2 is phytotoxic. Plant roots absorb O_2 for respiration and release CO_2. Soil microorganisms require O_2 for respiration and metabolism. Respiration creates an imbalance between these two gases. The gaseous exchange processes minimize this difference and establish a favorable balance between them. Restricted gaseous exchange leads to anaerobic conditions in soil, which are harmful for healthy plant growth and yield. Many undesirable physiological changes occur in plants, and many vital metabolic processes in soil and plant systems are adversely affected. Several microbial activities essential for a productive soil, such as decomposition of organic matter, nitrification, and sulfur oxidation, are retarded, while the production of methane (CH_4), which is a potent atmospheric pollutant and responsible for global warming, is enhanced. For normal performance, the growing plants should not experience O_2 stress and/or CO_2 excess at any of their physiological growth stages. A soil is said to be well aerated if it can maintain optimum O_2 supply for root respiration, microbial activities, and various electro- and biochemical as well as metabolic processes occurring in soil and plant system.

In this chapter, we shall discuss the mechanisms of soil aeration, evaluation and characterization of soil air, significance of soil air in plant growth, and agricultural practices to ensure favorable soil aeration status.

P. K. Sharma, S. Kumar, *Soil Physical Environment and Plant Growth*,
https://doi.org/10.1007/978-3-031-28057-3_6

6.1 Soil Aeration

Soil aeration refers to "the exchange of gaseous between soil and outer atmosphere." The two major gases associated with soil aeration are O_2 and CO_2. The O_2 moves from the atmosphere to soil, while CO_2 moves from soil to the atmosphere. The soil aeration, therefore, may also be defined as "the exchange of O_2 and CO_2 gases between soil and outer atmosphere." Soil aeration is a vital process as it maintains a balance between the two life-sustaining gases (O_2 and CO_2) in soil air. Oxygen is consumed (respiration of roots and microbial activities), while CO_2 is produced in soil (product of respiration). A proper balance of these two gases in the root zone is essential for a healthy plant growth.

6.2 Composition of Soil Air

The *composition of soil air* refers to "the relative concentration of gaseous components of soil air." The average composition of air in a well-aerated soil and the outer atmosphere is shown in Table 6.1.

The gases present in the outer atmosphere may be grouped into two categories:

i. *Permanent gases*, i.e., the gases whose concentration remains almost constant with time
ii. *Variable gases*, i.e., gases whose concentration keeps on changing with time

The 99% of the atmospheric air is composed of permanent gases and <1% of variable gases. The N_2 (78%) and O_2 (21%) are the two major components of permanent gases. There are several other permanent gases, such as argon, neon, helium, krypton, and hydrogen, but they are present in traces. The variable gases include water vapors, CO_2, CH_4, O_3, and N_2O.

The composition of soil air is quite similar to that of atmospheric air, except only in terms of CO_2 and water vapors, which are higher in soil air (Table 6.1). The composition of soil air is highly variable. It depends on factors like time of the year, soil temperature, soil moisture, soil organic matter status, soil reaction, soil depth, respiration activity of plant roots and other soil flora and fauna, solubility of O_2 and CO_2 in water, and gaseous exchange processes.

Table 6.1 Average composition of soil and atmospheric air

Component	Soil air (%)	Outer atmospheric air (%)
Oxygen	20.6	21.0
Carbon dioxide	0.25	0.03
Nitrogen	78.2	78.0
Relative humidity (water vapors)	95–100	Highly variable (20–90)

Adapted from Russell and Appleyard (1915)

6.3 Oxygen Requirement of Plants

Plants require some minimum amount of O_2 for their optimum growth and yield, and the requirement varies with plant species. Depending on the O_2 requirement of plants, the O_2 concentration in soil air may be divided into three categories:

i. *The optimum range:* the range of O_2 concentration over which plant growth is normal
ii. *The limiting value:* the O_2 concentration below which the plant growth starts declining
iii. *The critical value:* the O_2 concentration at which the plant growth stops completely

It has long been established that although some plants are able to grow even at O_2 concentrations as low as 0.5%, most plants stop growing at 2% O_2 content (Cannon and Free, 1925). Some aquatic/semiaquatic plant species, such as rice, have the capability to adapt to O_2-deficient conditions by mechanisms like developing aerenchyma or by hormonal adjustment (Armstrong et al. 1994). Rice can grow even in the complete absence of O_2 for a prolonged period.

To maintain O_2 in the root zone in the optimum range, the net O_2 flux into the soil surface should be around 10 L/m^2/d during summers and around 1.0 L/m^2/d during winters. Stolzy and Latey (1964) extensively reviewed the relationship between ODR and root/shoot growth of various plant species and concluded that at ODR of 20×10^{-8} g/cm^2-min, the roots of most of the plants cease to grow; at ODR between 20 and 30×10^{-8} g/cm^2-min, root growth is retarded; and an ODR of 50–70×10^{-8} g/cm^2-min is needed for the good emergence of most plant species. For the normal growth of most of the plants, the ODR should be $>40 \times 10^{-8}$ g/cm^2-min. The *limiting* and *critical values* of ODR below which seedling emergence declines rapidly and stops completely for some crops are shown in Table 6.2.

If O_2 supply falls short of its rate of consumption by plant roots, aerobic soil organisms, and biochemical reactions, the anaerobic conditions develop in soils, and the process is called *soil anaerobiosis*. Soil anaerobiosis refers to oxygen-free soil environment. The opposite of this is *soil aerobiosis*, i.e., oxygen-rich soil

Table 6.2 Limiting and critical values of ODR for some crops

	ODR (g/cm^2/min)	
Crops	Limiting value	Critical value
Barley	15×10^{-8}	5×10^{-8}
Oats	18×10^{-8}	7×10^{-8}
Beans	20×10^{-8}	7×10^{-8}
Wheat	24×10^{-8}	5×10^{-8}
Maize	24×10^{-8}	10×10^{-8}
Tomato	24×10^{-8}	15×10^{-8}
Sugar beet	30×10^{-8}	8×10^{-8}

Adapted from Glinski and Stepniewski (1985)

environment. Anaerobic conditions develop in soils due to waterlogging, poor drainage, shallow water table, amendment of soils with heavy rates of organic materials like animal wastes, composts and biosolids, or use of ammoniacal fertilizers. The soil anaerobiosis may be temporary (for a few hours or days) as in uplands and permanent (for months or years) as in wetlands. Depending on the severity and duration of causes of anaerobiosis, the O_2 content in soil may be deficient (a situation called *hypoxia*) or absent (a situation called *anoxia*). The impact of soil anaerobiosis on soil properties has been detailed in Chap. 4 (Sect. 4.4).

6.4 Gaseous Exchange Mechanisms

The exchange of gases between respiring sites (such as plant roots, microbes) and outer atmosphere takes place in two steps:

i. Gaseous exchange between outer atmosphere and air-filled soil pores
ii. Gaseous exchange between air-filled pores and respiring sites

The first process occurs primarily in gaseous phase, while the second process occurs through water films of varying thickness surrounding the living tissues (i.e., the respiring sites).

Gaseous exchange takes place by two mechanisms, viz. *mass flow* and *diffusion*. Gaseous diffusion dominates the mass flow process. The *rainfall* also plays an important role in supplying oxygen to soil.

6.4.1 *Mass Flow*

Mass flow, also known as *convective flow*, refers to “the exchange of air between outer atmosphere and soil in response to the total air pressure gradient.” The total air pressure gradients between soil and atmosphere arise due to changes in meteorological factors such as rain, wind, temperature, and barometric pressure. The total air pressure gradient acts as the driving force for the flow of air. Mass flow causes movement of the entire air mass. Rainwater (or irrigation) displaces air in the soil pores during infiltration and also brings dissolved O_2 in soil. The O_2-saturated water at atmospheric pressure contains around 6 mL of O_2 per liter.

6.4.2 *Diffusion*

Diffusion is the most important mechanism of gaseous exchange between atmosphere and soil pores, and between soil pores and respiring sites (i.e., roots and micro and macro soil fauna). *Diffusion* of soil air refers to “the movement of gases through diffusion in response to their partial pressure gradients, i.e., concentration gradient of individual gases." The concentration gradients of gases in soil may be established due to several processes such as:

- Respiration of microorganisms, macrofauna, and plant roots
- Production of gases due to biological reactions such as fermentation, nitrification, and denitrification
- Incorporation of chemicals such as fumigants, anhydrous ammonia, pesticides, and volatile organic compounds from toxic waste sites

Gaseous diffusion rates through water films are much slower than in air (Sect. 4.4.2).

6.5 Evaluation of Soil Air

Soil air may be evaluated by measuring the total gaseous content in soil pores or O_2 content in soil air or some soil property which is related to soil aeration status. Several indices have been developed for the characterization of soil air, and the choice depends on the objective and accuracy of measurement. Important indices of soil aeration are briefly described below:

6.5.1 Air-Filled Porosity

The *air-filled porosity* (also called *air capacity*) is a measure of the soil pore space filled with air. The air-filled porosity may be defined as "the volume fraction of pore space filled with air." Mathematically,

$$f_a = f - \theta \tag{6.1}$$

$$\text{Also } f_a = V_a / V_t \tag{6.2}$$

where f_a is air-filled porosity (%), f the total porosity (%), θ the volume wetness (%), V_a the volume of air-filled pores (cm^3), and V_t the bulk volume of soil (cm^3).

The air-filled porosity varies with the soil water content. It is important, therefore, to express air-filled porosity along with the soil water content. Different soils should be compared for aeration status at the same water content. The air capacity expressed at field capacity is called *field air capacity*. For normal plant growth, the air-filled porosity should be >10%. This index gives the current aeration status of soil but fails to consider aeration dynamics.

Air-filled porosity depends on soil texture, structure, drainage conditions, water content, and organic matter status. Sandy soils, aggregated soils, well-drained soils, and OM-rich soils generally have higher air-filled porosity at a given soil water content. The critical value of 10% air-filled porosity in these soils generally arrives early after rainfall/irrigation and at relatively higher water contents. In a long-term field experiment under maize-wheat system in a silty clay loam soil, Verma and Sharma (2008) found 10% air-filled porosity at 39.2%, 37.2%, and 35.5% volume wetness, respectively, in plots treated with (NPK+FYM), NPK, and untreated control. Further, the 10% air-filled porosity was reached earlier in (NPK+FYM) and NPK plots, i.e., in 5 and 8 days after irrigation, respectively, as compared to 12 days after irrigation in control plots.

The air-filled porosity (f_a) is either measured directly by using air pycnometer method or calculated from total porosity (f) and volume wetness (θ) using Eq. 6.1. Total porosity may be computed from particle density (ρ_s) and bulk density (ρ_b) values of soil using the following equation:

$$f = 1 - \rho_b / \rho_s \tag{6.3}$$

6.5.2 *Aeration Porosity*

The *aeration porosity* is defined as "the pore space filled with air when soil is under a suction of 50 cm water column." It is also known as *aeration capacity* or *non-capillary porosity*. At 50 cm suction, soil pores ≥0.06 mm diameter (i.e., non-capillary pores) are free of water and contain air. For normal plant growth, the aeration porosity should be ≥10%.

The aeration porosity largely depends on soil texture and structure. Aeration porosity may be around ≥25% in sandy soils, 15–20% in loam soils, and ≤10% in clayey soils. Aggregated soils and soils with crumb structure generally have higher aeration porosity (20–30%) than poorly aggregated or compacted soils (≤5%).

The aeration porosity, like the air-filled porosity, indicates the existing aeration status but fails to consider the aeration dynamics. It is a simple index and is determined with pressure plate apparatus or hanging water column technique.

6.5.3 *Composition of Soil Air*

The O_2 and CO_2 are the two important constituents of soil air in relation to plant growth. Their concentration in soil air serves as an index of soil aeration. The average O_2 content of soil air in well-aerated soils is around 20.6% (Table 6.1). In waterlogged soils, the O_2 concentration may drop to 0%. The O_2 concentration >10% is required for the normal growth of most of the plants (Kohnke, 1968; Kramer, 1969).

The average CO_2 content of soil air is around 0.25%. The CO_2 at higher concentrations is toxic to plants. Cultivated soils, due to respiration by roots and soil flora and fauna, generally have lesser O_2 and more CO_2 than in fallow lands. Further, CO_2 concentration is higher near plant roots than farther away due to root respiration. Higher CO_2 contents in soil air adversely affect root growth, root water absorption, chlorophyll, starch content, and total biomass; soil microbial acid phosphatase activity is retarded (He et al. 2019).

The gas composition of air samples may be determined with the help of infrared gas analyzers, oxygen analyzers, or gas chromatographs. It is possible to analyze 1 mL gas sample very accurately with gas chromatography. Knowing the composition of soil air is more reliable than measuring air volume alone (i.e., air-filled porosity or aeration porosity). A combination of air volume and air composition is a better indication of soil aeration status. The air composition as an index of soil aeration has the disadvantage of being a static measurement. Taking representative air samples is also a problem.

6.5.4 Air Permeability

Air permeability is defined as "the coefficient governing the mass flux of air through soil in response to total pressure gradient." In simple words, the air permeability indicates as to how fast air can pass through soil mass under a given air pressure condition. Air permeability is a measure of the openness and continuity of soil pores because the flow of gases in soil depends on the size and continuity of air-filled pores. Higher air permeability is an indicator of better aeration status. Air permeability is also a sensitive index of soil structure as it is highly responsive to soil structural modifications.

The air permeability can be measured in two ways: (i) the constant pressure head method and (ii) the falling-pressure head method. In *constant pressure method*, which is a steady-state method, the volume of gas flowing through a soil sample of known surface area under constant gas pressure is measured as a function of time. In *falling-pressure head method*, the gas is allowed to pass through a soil sample and the change in gas pressure in a constant volume is measured, which is an index of air permeability.

6.5.5 Oxygen Diffusion Rate (ODR)

For normal plant growth, the rate of O_2 supply is more important than the absolute concentration in the soil at any given time. The rate of O_2 supply, thus, is a sensitive index of soil aeration. The O_2 supply rate is determined as oxygen diffusion rate.

The *oxygen diffusion rate* (ODR) is an index of O_2-supplying power of soils to growing plant roots. The ODR is defined as "the rate at which oxygen diffuses to the plant roots in soil." Studies have shown that root growth of most of the plants ceases at an ODR $\leq 20 \times 10^{-8}$ g/cm^2-min; root growth is retarded at an ODR between 20 and 30×10^{-8} g/cm^2-min; and for normal root and shoot growth, the ODR should be $\geq 40 \times 10^{-8}$ g/cm^2-min (Stolzy and Latey, 1964).

Different methods are available for determining ODR in soils. The most commonly used method is the *platinum electrode* technique given by Lemon and Erickson (1952). The platinum electrode technique of measuring ODR is based on the *principle* that the reduction of O_2 molecules at the platinum electrode surface (which simulates the plant root) produces electric current, which is governed solely by the rate at which O_2 diffuses to the electrode surface through water film surrounding the electrode. The measurement of resulting current is a measure of ODR in soil.

The electric current flowing between two electrodes (i) and the oxygen flux (q_o) at the electrode surface are related as per the following equation:

$$i = nFAq_{o,t} \tag{6.4}$$

$$\text{or} \quad q_{o,t} = i / nFA \tag{6.5}$$

where $q_{o,t}$ is the oxygen flux at the surface of electrode, i.e., the oxygen diffusion rate (moles/cm^2-min) at time t; i the electric current (microamperes, μA); n the number of electrons required to reduce one molecule of oxygen (n = 4); F the Faraday constant (96,500 coulombs/mole of oxygen); and A the surface area of the electrode (cm^2).

For calculating O_2 flux ($q_{o,t}$), the steady-state current (i) is measured after 4–5 min, with the assumption that the rate of O_2 reduction depends on O_2 diffusion rate, and that both are equal. By using these values, ODR can be calculated as follows:

$$ODR = q_{o,t} = iM / nFA \tag{6.6}$$

where M is the molecular weight of oxygen (32 g/mole). Substituting values of M, n, and F in Eq. 6.6, the ODR becomes equal to

$$ODR\left(\text{g}/\text{cm}^2 - \text{min}\right) = \left(60 \times 32 \times i\right) / \left(4 \times 96500\right) \times A \tag{6.7}$$

$$\text{or} \quad ODR\left(\text{g}/\text{cm}^2 - \text{min}\right) = 0.00497\left(i / A\right) \tag{6.8}$$

The factor 60 is used to convert time from seconds to minutes. It is because one ampere is equal to one coulomb per second, i.e., 1.0 coulomb/s = 1.0 ampere.

If platinum microelectrode is having 4 mm length and 0.65 mm diameter, Eq. 6.8 changes to

$$ODR = 0.059i\left(\text{g}/\text{cm}^2 - \text{min}\right) \tag{6.9}$$

6.5.5.1 Factors Affecting ODR

The ODR in soils is influenced collectively by a number of factors such as:

- Bulk density, i.e., total soil porosity
- Volume of air-filled pores
- O_2 content of soil air and soil solution
- O_2 diffusion coefficient in soils
- Air permeability of soils
- Thickness of water film around platinum electrode
- Length of O_2 diffusion path to the electrode surface
- Radius (i.e., surface area) of the electrode
- Soil temperature, with every one-degree Celsius (°C) rise in temperature, the ODR increasing by 1.4–2.4%

6.5.5.2 Limitations of ODR

The ODR technique has certain limitations:

- *Applicable in wet soils*: The platinum electrode technique is applicable in wet soils and not in moist to dry soils. It is because the electrode must remain moist, i.e., covered with water film during its use.
- *Electrode poisoning*: Keeping the platinum electrodes permanently embedded in soil causes deposition of several types of chemicals, such as iron, copper, aluminum, sulfur compounds, calcium carbonates, phosphates, arsenic, and organic compounds, on the surface of platinum surface. This process is called *electrode poisoning*. The poisoned electrodes measure ODR values lower than the freshly inserted electrodes. The reduction in measured values may be up to 50–60%. Hence, it is advisable to insert electrodes in soil at the time of making ODR measurements, and not keeping them permanently embedded in soil for the entire crop season or so.
- *Relatively long time needed for taking measurement*: For steady-state current measurement, the current should be measured after 4–5 min of equilibrium of platinum electrode in soil.

Numerical Problem

Example I. Calculate ODR for two measured current values, i.e., 0.5 and 5 microamperes, if the length of the platinum microelectrode is 4 mm and diameter is 0.6 mm. Based on the ODR values, give your comments on soil

aeration status with respect to crop plants. *Solution*: Given: i = (a) 0.5 μA and (b) 5 μA

$$A = r(2h + r)\pi$$

$$r = 0.6/2\,\text{mm} = 0.3\,\text{mm} = 0.03\,\text{cm}$$

$$h = 4\,\text{mm} = 0.4\,\text{cm}$$

$$\therefore \quad A = 0.03\{(2\times 0.4) + 0.03\}22/7$$

$$A = 0.07826\,\text{cm}^2$$

From Eq. 6.8,

$$ODR = 0.00497(i/A)\,\mu\text{g}/\text{cm}^2 - \text{min}$$

$$\text{Case}(\text{a}) \quad ODR = 0.00497(0.5/0.07826)$$

$$= 0.32\times 10^{-8}\,\text{g}/\text{cm}^2 - \text{min}$$

$$\text{Case}(\text{b}) \quad ODR = 0.00497(5/0.07826)$$

$$= 31.8\times 10^{-8}\,\text{g}/\text{cm}^2 - \text{min}$$

The ODR in case (a) is below the critical value, indicating poor aeration not suited for cultivation of arable crops; in case (b), the ODR is above the limiting value, indicating good aeration status suitable for cultivation of almost all arable crops.

6.5.6 Soil Redox Potential

Soil redox potential is a valuable tool to determine the O_2 status of wet soils (waterlogged soils or saturated soils). In wet soils, elemental O_2 is almost zero, and ODR may also be zero or too low. The *oxidation-reduction potential*, also called the *redox potential* (Eh), is a measure of the intensity of reduction in soil in the absence of O_2. The redox potential determines the tendency of a system to oxidize or reduce the chemical substances. The oxidation-reduction is a chemical reaction in which electrons are transferred from a donor to an acceptor. Loss of electrons is "oxidation," while gain of electrons is "reduction." High or positive Eh values indicate oxidized condition, while low or negative values indicate reduced condition. The range of Eh is much wider, varying approximately between −300 and +700 mV. Oxygen becomes absent at Eh values of ≤+350 mV. The impact of waterlogging on soil redox potential is discussed in Chap. 4 (Sect. 4.4.3).

The Eh can be measured with the help of a platinum electrode method.

6.6 Comparison of Different Indices of Soil Aeration

Each index of soil aeration has its own merits and demerits in terms of operational feasibility, accuracy, purpose of measurement, collection of representative air samples, etc. Some techniques require expensive equipment, which necessitates time-to-time standardization and handling skills; some provide static information; and some are difficult to employ routinely. Studies have shown that under fluctuating soil moisture conditions, a frequent record of air-filled porosity serves as a better index than ODR or O_2 content of soil air. The ODR is a sensitive index in arable lands, while Eh is useful in wetlands. It is advisable to use those indices for routine field studies which are simple and can be measured easily and frequently.

6.7 Soil Air and Plant Growth

Soil aeration is an important factor in the normal growth of plants. The O_2 and CO_2 in the root zone must be in proper balance. For that, adequate supply of O_2 to plant roots and rapid removal of CO_2 from the soil atmosphere are very essential. The poor soil aeration affects plant growth in several ways as briefly discussed below:

6.7.1 Root and Shoot Growth

The immediate effect of reduction in O_2 content and increase in CO_2 content in the soil air is reduction in root respiration. Scotter et al. (1967) reported that a reduction in O_2 concentration around the roots of dwarf peas from 21% to 10%, 5%, and 2.5% caused a reduction in root respiration by 15%, 40%, and 70%, respectively.

Energy is released during aerobic respiration through glucose–oxygen reaction. One molecule of glucose ($C_6H_{12}O_6$) combines with six molecules of O_2 to produce 38 adenosine triphosphate (ATP) molecules as per the following reaction:

$$\underset{\text{(Glucose)}}{C_6H_{12}O_6} + 6O_2 \rightarrow 6CO_2 + 6H_2O + \underset{\text{(Energy)}}{38\,\text{ATPs}} \tag{6.10}$$

The ATPs are the energy molecules. Reduction in root respiration reduces energy production, which adversely impacts vital plant processes, including metabolic processes within the plant system.

The first observable effect of poor soil aeration is wilting of the plant. The lower leaves generally show first signs of wilting. If the poor aeration conditions continue, the wilting extends over the whole plant, leading to leaves turning brown or yellow and which eventually curl up and drop. The speed at which the wilting occurs and leaves turn yellow and drop depends on the duration of aeration stress and the type of the crop.

The root development is severely affected. Root growth is reduced in size and form. Root growth rate is either retarded or completely stopped depending on the severity of aeration stress. Roots become stunted and abnormally shaped. Root branching is limited, and root hairs are drastically reduced. Root permeability is reduced. Such physiological modifications in plant roots decrease their capacity to absorb moisture and nutrients from the soil and their translocation within the plant system. The "root growth pressure" is also relatively low in poorly aerated soils. It enhances the adverse effects of soil compaction on root growth and development.

Root and shoot growth is thus retarded due to poor soil aeration, which leads to poor biomass production and crop yield. The prolonged aeration stress may even lead to death of plants.

6.7.2 Soil Microflora and Fauna

All microflora and fauna living in the soil require O_2 for respiration and metabolism. Their population is drastically reduced in O_2-deficient soils. The accumulation of CO_2 in soil air further aggravates the problem. The microbial activities are slowed down (He et al. 2019), leading to reduction in decomposition rates of organic matter, nitrification, sulfur oxidation, etc. These processes adversely impact plant growth.

6.7.3 Accumulation of Phytotoxins

The soil anaerobiosis leads to reduced soil conditions which favor production and accumulation of several organic and inorganic phytotoxins, such as organic acids (acetic acid, butyric acid, phenolic acid, etc.), hydrogen sulfide, reduced iron and manganese compounds, ferrous sulfide, nitrites, methane, and ethylene, in soil (Sect. 4.4.3). These toxins have an adverse effect on plant growth and yield.

6.7.4 Water and Nutrient Absorption

The O_2 deficiency retards respiration, which has a negative effect on water and nutrient absorption by the plants. The effects become more severe with the accumulation of CO_2 in soil air (Free, 1917; Kramer, 1940; Chang and Loomis, 1945). Plants use energy produced during respiration in the absorption of water and nutrients. Such a situation may arise in waterlogged soils.

6.7.5 Development of Plant Diseases

Insufficient aeration increases the susceptibility of plants to pathogens and diseases, particularly to root-rotting fungi (Louvet, 2020). Poor soil aeration favors the growth of water molds like *Pythium* and *Phytophthora*. Appearance of diseases like wilt of gram and dieback of citrus and peach are associated with O_2 stress in the root zone. It is also observed that in poorly aerated soils, even weak pathogens which normally do not harm plants in aerated soils may become extremely destructive.

6.8 Management of Soil Air

The aeration management strategies aim at controlling factors which are responsible for soil aeration problem. It is important, therefore, to understand first the causes of poor soil aeration before aeration management strategies are decided.

6.8.1 Causes of Poor Soil Aeration

Soil aeration problems may arise due to one or more of the following reasons:

- *Waterlogging*: Waterlogging, temporary or permanent, is the major cause of poor soil aeration. Excess water replaces almost all the air from soil pore spaces. Whatever little O_2 is left (may be in soil solution) is rapidly consumed in respiration processes with the release of CO_2. Consequently, O_2 content drops and imbalance is created between O_2 and CO_2. Diffusion rates of O_2 and CO_2 through water films are around 1/10,000 times that in air. Gaseous diffusion in waterlogged soils is, therefore, drastically reduced.
- *Excess rainfall or over-irrigation:* Soils may get saturated or nearly saturated temporarily due to excess rainfall or over-irrigation. Ponding of water may also occur in depressions over otherwise flatlands. It leads to aeration problems. Such situations are frequently encountered in high-rainfall areas.

- *Shallow water table:* Water tables close to the soil surface keep the root zone saturated with water through capillary action. Shallow water tables also lower the deep percolation of water, leading to ponded water conditions.
- *Low water-permeable soils:* Soils having low infiltrability and saturated hydraulic conductivity usually suffer from aeration problems. Such situations are generally encountered in fine-textured and poorly structured soils.
- *Soil compaction:* Surface or subsurface soil compaction, caused by natural processes (such as illuviation of clay) or anthropogenic activities (such as the use of heavy farm machines), makes agricultural lands prone to aeration problems. Soil compaction decreases soil porosity and saturated hydraulic conductivity of soils and facilitates water ponding at soil surface. It is a problem of mechanized farming.
- *Soil trampling by livestock:* Animal grazing may cause compaction of pasturelands due to their hoofs if the animals stay long in the same area. The problem is more severe in wetlands.
- *Applications of organic matter:* Frequent applications of organic matter to soils in huge amounts may lead to soil aeration problems. Decomposition of organic matter releases CO_2. The rate of decomposition of organic matter and, thus, the release of CO_2 in soils are proportional to the soil organic matter content. If CO_2 is not exchanged with the atmosphere rapidly, its concentration in soil air may rise to toxic levels, with simultaneous decline in O_2 content.

6.8.2 Management of Poor Soil Aeration

Once the cause of problem of soil aeration is known, it becomes easy to select appropriate management strategy(ies). Different management practices have been developed to control different factors causing aeration problem. It is always advisable to adopt a combination of technologies for better results. Also, steps should be taken to improve soil aeration as soon as the plants start showing aeration stress symptoms like wilting of leaves or reduction in vigor of growth, number of blooms, and size of blooms. Steps taken at advanced stage may lead to yield and monetary loss. Following are some of the practices for improving soil aeration conditions:

- *Improve soil drainage*: Adopt surface or subsurface drainage systems to improve soil drainage (Sect. 4.5).
- *Restricted use of farm machinery*: If aeration problem is caused by soil compaction due to heavy farm machinery use, limit the movement of heavy machines and reduce their frequency on farmlands. Operate machines at optimum soil moisture conditions when soil is in friable consistency.
- *Rotational pastures*: Compaction of pasturelands caused by trampling of livestock may be avoided or minimized by rotating the pastures for grazing. Avoid animal grazing when soils are wet.

- *Moderate use of organic matter*: Use organic matter in moderate amounts or in decomposed state to reduce decomposition processes consuming O_2 and releasing CO_2. Applications of organic matter otherwise are very useful and essential to improve soil structure, and thus soil aeration. Mulching and use of organic materials such as compost, FYM, and organic farm residues or use of soil amendments is recommended for improving soil structure and associated soil properties. The technology is particularly suited to clay soils, which are more subject to aeration problems. Addition of organic matter also activates soil organisms, like earthworms, which make soil more porous.
- *Tillage:* Tillage operations (e.g., digging, pulverization, turning of soil upside down, deep tillage) loosen the soil, enhance soil porosity, improve hydraulic properties of soil, and thus improve soil aeration conditions. Tillage practices are also important components of "integrated weed control" and "integrated pest/disease control." Tillage is effective in improving soil aeration, but its effects are short lived. It also has the disadvantage of burning fossil fuel and generating greenhouse gases.
- *Spike/core/liquid aeration*: Soil aeration can also be improved by techniques like spike aeration, core aeration, and liquid aeration.

- *Spike aeration* is done by punching holes on the soil surface to facilitate the entry of air into soil. Spiking is done with tools like soil aeration shoes, prongs, rollers, and mower attachments. The technique is good for sandy soils but not suitable for clay soils.

- *Core aeration* involves digging out small soil cores at soil surface with manual aerators or mower attachments and leaving the cores there on the soil surface. This technique is useful in clay soils.

- *Liquid aeration* is done to improve biological activities in soil. In this technology, liquid aerators (e.g., ClayMend) containing beneficial soil microbes, nutrients for soil biota, and wetting agents are injected into soil. They stimulate soil organisms to dig deeper in soil to make it porous and improve soil aeration. Such products are commercially available.

6.9 Question Bank

6.9.1 Short Questions:

(i) Differentiate between soil air and soil aeration. What is the composition of soil air in comparison to air in the outer atmosphere?
(ii) What is the oxygen requirement of plants? In what way oxygen stress impacts the plant growth?
(iii) What are different indices of characterizing soil air? Which index/indices is/are commonly used in upland soils and why?

(iv) Differentiate between aeration porosity and air-filled porosity. How are these indices determined?
(v) Briefly explain "oxygen diffusion rate" (ODR) as an index of soil aeration. What is electrode poisoning?
(vi) How does soil aeration affect root and shoot growth of plants?
(vii) Explain the factors responsible for poor soil aeration. How will you manage soils with poor aeration?
(viii) What is soil anaerobiosis? How does it develop in soils?
(ix) How does gaseous exchange occurs in soils? Briefly explain mass flow and diffusion processes.

6.9.2 *Briefly Explain Why?*

(i) The concentration of CO_2 is generally higher in soil air than in outer atmospheric air.
(ii) Aquatic plants can grow in O_2-deficient soils.
(iii) The platinum electrodes permanently embedded in soil may underestimate ODR than freshly inserted electrodes.
(iv) The platinum electrode technique of measuring ODR is not applicable in moist to dry soils.
(v) Plant appearance may give some indication of poor soil aeration.
(vi) Plants perform poor in poorly drained soils.
(vii) Waterlogging is the major cause of poor soil aeration.
(viii) Soils with shallow table generally have aeration problems.
(ix) Grazing of pastures may also impact the aeration status of pasturelands.
(x) Conservation tillage improves and sustains good soil aeration.

6.9.3 *Fill in the Blanks:*

(i) The process of air exchange between soil and outer atmosphere is known as …….
(ii) The ………….. is defined as "the pore space filled with air when soil is under a suction of 50 cm water column."
(iii) The concentration of ……….. and ……….. is almost the same in soil air and outer atmosphere.
(iv) The concentration of …….. is higher in soil air than in outer atmospheric air.
(v) The roots of most of the plants cease to grow at or below ODR of ……….. g/cm^2-min.
(vi) For the normal growth of most of the plants, the ODR should be ………….. g/cm^2-min.

(vii) The O_2 supply to soil at rates short of consumption by plant roots leads to anaerobic conditions, a process called ....
(viii) In mass flow of gases in soils, the driving force is ....
(ix) In diffusion of gases in soils, the driving force is .... of gases.
(x) The volume fraction of pore space filled with air at any given soil water content is called ....
(xi) For normal plant growth, the air-filled porosity should be ....
(xii) The rate at which oxygen diffuses to the plant roots in soil is called ....
(xiii) Oxygen becomes absent in soils at Eh values of ....
(xiv) The "root growth pressure" is relatively .... in poorly aerated soils.
(xv) Diffusion rates of O_2 and CO_2 through water films are around .... times that in air.
(xvi) The O_2 concentration of soil air at which the plant growth stops completely is called the ....

[**Key:** i. soil aeration, ii. aeration porosity, iii. oxygen and nitrogen, iv. carbon dioxide *or* water vapors, v. 20×10^{-8}, vi. $>40 \times 10^{-8}$, vii. soil anaerobiosis, viii. total air pressure gradient, ix. partial pressure gradient, x. air-filled porosity, xi. >10%, xii. oxygen diffusion rate, xiii. ≤+350 mV, xiv. low, xv. 1/10,000, xvi. critical value]

6.9.4 State Whether the Following Statements Are True (T) or False (F):

(i) The poisoned platinum electrodes overestimate ODR.
(ii) The O_2 deficiency retards respiration, which has a negative effect on water and nutrient absorption by the plants.
(iii) Weak pathogens which normally do not harm plants in aerated soils may become extremely destructive in poorly aerated soils.
(iv) The oxygen stress in soil generally suppresses diseases like wilt of gram and dieback of citrus plants.
(v) Waterlogging is the major cause of poor soil aeration.
(vi) Low water-permeable soils have high oxygen concentration in soil air.
(vii) Animal grazing may cause compaction of pasturelands and create aeration problem.
(viii) Frequent applications of organic matter to soils in huge amounts may lead to soil aeration problems.
(ix) Subsurface drainage lowers soil water content but has no effect on O_2 concentration of soil air.
(x) Spike aeration is a process of improving soil aeration.
(xi) Soil air and soil aeration are synonymous terms.
(xii) Air-filled porosity and aeration porosity are synonymous terms.
(xiii) The composition of soil air is quite similar to that of atmospheric air, except only in terms of CO_2 and water vapors.

(xiv) The limiting value of O_2 concentration in soil air is one below which the plant growth stops completely.

(xv) The critical value of O_2 concentration in soil air is one below which the plant growth starts declining.

(xvi) Most crop plants stop growing at 2% O_2 content.

(xvii) For the normal growth of most of the plants, the ODR should be $>40 \times 10^{-8}$ g/cm^2-min.

(xviii) Wheat is more sensitive to oxygen stress than barley.

(xix) Mass flow is the major process of gaseous exchange in soils.

(xx) Plant growth is adversely affected if air-filled porosity falls below 10%.

[**Key:** i. F, ii. T, iii. T, iv. F, v. T, vi. F, vii. T, viii. T, ix. F, x. T, xi. F, xii. F, xiii. T, xiv. F, xv. F, xvi. T, xvii. T, xviii. T, xix. F, xx. T]

6.9.5 Multiple-Choice Questions:

(i) The composition of soil air is quite similar to that of atmospheric air except in terms of

(a) Oxygen
(b) Nitrogen
(c) Carbon dioxide
(d) Carbon dioxide and water vapors

(ii) The critical value of oxygen requirement of plants is one

(a) Over which plant growth is normal
(b) Below which the plant growth starts declining
(c) At which the plant growth stops completely
(d) Below which biomass production reduces by 50%

(iii) The critical value for most of the plants is

(a) 0% oxygen content in soil air
(b) 2% oxygen content in soil air
(c) 10% oxygen content in soil air
(d) 20% oxygen content in soil air

(iv) To maintain O_2 in the root zone in the optimum range, the net O_2 flux into the soil during summer surface should be

(a) Around 10 L/m^2/d
(b) Around 20 L/m^2/d
(c) Around 30 L/m^2/d
(d) Around 40 L/m^2/d

(v) For normal plant growth of most of the crop plants, the oxygen diffusion rate should be

(a) $>10 \times 10^{-8}$ g/cm^2-min
(b) $>20 \times 10^{-8}$ g/cm^2-min
(c) $>30 \times 10^{-8}$ g/cm^2-min
(d) $>40 \times 10^{-8}$ g/cm^2-min

(vi) Aerobiosis refers to

(a) O_2-rich environment
(b) CO_2-rich environment
(c) Air exchange between soil and outer atmosphere
(d) Air requirement of growing plant roots

(vii) The driving force in mass flow of air in soil is

(a) Total air pressure gradient
(b) Partial pressure gradient
(c) Air content in soil pores
(d) Atmospheric evaporativity

(viii) The driving force in the diffusion of gases in soil is

(a) Total air pressure gradient
(b) Partial pressure gradient of the gas
(c) Air content in soil pores
(d) Atmospheric evaporativity

(ix) For normal plant growth, the air-filled porosity should be

(a) >10%
(b) >20%
(c) >30%
(d) >40%

(x) The major constraint to plant growth in waterlogged soils is

(a) O_2 stress
(b) CO_2 toxicity
(c) A combination of O_2 stress and CO_2 toxicity
(d) Excess of water content

(xi) Oxygen becomes absent in soil air at Eh values of

(a) ≤ -700 mV
(b) ≤ -350 mV
(c) ≤ -100 mV
(d) $\leq +350$ mV

[**Key:** i. d, ii. c, iii. b, iv. a, v. d, vi. a, vii. a, viii. b, ix. a, x. c, xi. d]

References

Armstrong W, Braendle R, Jackson M (1994) Mechanisms of flood tolerance in plants. Acta Botanica Neerlandica 43:307–358

Cannon WA, Free EE (1925) Physiological features of roots with especial references to the relation of roots to aeration of the soil. Carnegie Inst Publ:368

Chang HT, Loomis WE (1945) Effect of carbon dioxide on absorption of water and nutrients by roots. Plant Physiol 20(2):221–232

Free EE (1917) Effect of aeration on the growth of buckwheat in water cultures. Johns Hopkins Univ Cir 293:198–199

Glinski J, Stepniewski W (1985) Soil aeration and its role for plants. CRC Press, Boca Raton, FL

He W, Yoo G, Moonis M, Kim Y, Chen X (2019) Impact assessment of high soil CO_2 on plant growth and soil environment: a greenhouse study. Peer J 7:e6311

Kohnke H (1968) Soil physics. McGraw-Hill, New York

Kramer PJ (1940) Causes of decreased absorption of water by plants in poorly aerated media. Am J Bot 27:216–220

Kramer PJ (1969) Plant and soil water relationships—a modern synthesis. McGraw-Hill, New York

Lemon ER, Erickson AE (1952) The measurement of oxygen diffusion in the soil with a platinum electrode. Soil Sci Soc Am Proc 16:160–163

Louvet J. Effect of aeration and of concentration of carbon dioxide on the activity of plant pathogenic fungi in the soil. In: TA Toussoun, Robert V. Bega and Paul E. Nelson, editors. Root diseases and soil-borne pathogens. University of California Press 2020; 89–92

Russell EJ, Appleyard A (1915) The atmosphere of the soil, its composition and cause of variation. J Agr Sci 7:1–48

Scotter DR, Thurtell GW, Tanner CD (1967) Measuring oxygen uptake by the roots of intact plants under controlled conditions. Soil Sci 104:374–378

Stolzy LH, Latey J (1964) Characterizing soil oxygen conditions with a platinum microelectrode. Adv Agronomy 16:249–279

Verma S, Sharma PK (2008) Long-term effects of organics, fertilizers and cropping systems on soil physical productivity evaluated using a single value index (NLWR). Soil Tillage Res 98:1–10

Chapter 7
Soil Temperature and Plant Growth

Soil temperature is an important environmental factor affecting plant growth. All soil processes are temperature dependent. The soil thermal regime strongly influences the edaphic environment, which ensures crop productivity and sustainability. Soil temperature controls physical, chemical, physicochemical, biological, and biochemical processes in soil and plant systems (Onwuka and Mang 2018; Alli and Omofunmi 2021). In soil, it influences microbial and enzymatic activities, organic matter accumulation/decomposition and mineralization processes, water and nutrient availability to plants and microbes, soil aeration, humidity, fertilizer efficiency, pesticide degradation, etc. There is an approximately 2% change in various parameters present in the soil with every degree change in the temperature (Frey et al. 2013). In plants, it influences the seed germination, plant development and turgidity, water and nutrient uptake, and disease and insect occurrence. The rate of chemical and biochemical reactions doubles with every 10 °C rise in temperature. Soil acts as a store as well as a sink of thermal energy on a day-to-day or seasonal basis. It is important to understand soil thermal dynamics and their influence on plant growth in order to manage soils and crops in such a way to optimize soil temperature conditions.

In this chapter, we shall study the temperature requirement of plants in relation to plant growth, water and nutrient absorption by roots, and evaluation and management of soil temperature.

P. K. Sharma, S. Kumar, *Soil Physical Environment and Plant Growth*,
https://doi.org/10.1007/978-3-031-28057-3_7

7.1 Important Terminology

7.1.1 Heat

The *heat* may be defined as "the kinetic energy of the molecules of a substance due to their random motion." Heat can be transmitted through solids, fluids, gases, or vacuum.

7.1.2 Calorie

Calorie is a unit of heat in the International System of Units (SI). One *calorie* is "the amount of energy (heat) required to raise the temperature of one gram of pure liquid water by one degree Celsius."

7.1.3 British Thermal Unit

The *British thermal unit* (Btu) is a unit of heat in "foot-pound-second" (FPS) system. One Btu is "the amount of heat required to raise the temperature of one pound of pure liquid water by one degree Fahrenheit."

7.1.4 Temperature

Temperature may be defined as "the condition of a body which determines the transfer of heat to or from other bodies." It is a measure of the kinetic energy of the molecules of a substance, i.e., the intensity of heat. The temperature is expressed in degrees Fahrenheit (°F) or degrees Celsius (°C) or degrees Kelvin (°K).

7.1.5 Thermal Capacity

The *thermal capacity* (also called *heat capacity*) is defined as "the amount of heat that is present in a body." When expressed as "heat per unit mass," it is called *specific heat capacity* or *mass heat capacity* and measured as cal/g. When expressed as "heat per unit volume," it is called *volumetric heat capacity* and is measured as cal/cm^3. One gram-calorie (cal/g) is the amount of heat required to raise the temperature of 1 g of water from 15 to 16 °C.

7.1.6 Thermal Conductivity

Thermal conductivity is defined as "the quantity of heat transmitted through unit length of a substance per unit cross-sectional area, per unit of temperature gradient, per unit of time." It is the property of a substance to conduct heat. It is expressed as cal/s-cm-°C or J/m^2-s-°C in CGS system, Watt/m.°K in SI system, or BTU/hr.ft.°F in FPS system.

7.1.7 Thermal Diffusivity

The *thermal diffusivity* may be defined as "the change in temperature produced in a unit volume of substance by the quantity of heat flowing through it per unit time and per unit temperature gradient." It may also be defined as "the thermal conductivity per unit volumetric heat capacity." It is a measure of the ability of a substance to conduct thermal energy relative to its ability to store thermal energy. It indicates the rate of change of temperature with time. It may be expressed as cm^2/s (or m^2/s).

7.1.8 Thermal Retentivity

Thermal retentivity is the reciprocal of thermal diffusivity. *Thermal retentivity* may be defined as "the volumetric heat capacity per unit of thermal conductivity." It indicates the ability of the substance to retain heat and is expressed as s/cm^2 (or s/m^2).

7.1.9 Radiation

Radiation may be defined as "the energy emitted by a body above absolute zero temperature in the form of electromagnetic waves." Radiation is measured as heat energy per unit area per unit time (cal/cm^2-min). One cal/cm^2 is called one *langley.* Solar radiation is the primary source of heat in soil.

7.1.10 Shortwave Radiation

Shortwave radiation refers to the radiation having a wavelength between 0.3 and 4 μ (microns). Around 99% solar radiation is in shortwaves.

7.1.11 Long-Wave Radiation

Long-wave radiation refers to the radiation having a wavelength between 6.8 and 100 μ. Soil radiates heat mostly in long waves.

7.1.12 Solar Constant

Solar constant may be defined as "the rate at which the solar radiation is received at the top of the earth's atmosphere." It is estimated to be equal to 2.0 langleys per minute.

7.1.13 Diurnal Temperature Fluctuations

The *diurnal temperature fluctuations* refer to "the daily (24-h period) changes in soil temperature."

7.1.14 Seasonal/Annual Temperature Fluctuations

The *seasonal/annual temperature fluctuations* refer to "the seasonal or annual changes in soil temperature."

7.1.15 Amplitude

The *amplitude* may be defined as "the temperature difference between the average and peak (maximum or minimum) temperature." If the average is for a day, it is called *diurnal amplitude*, and if the average is for a season or year, it is called *seasonal/annual amplitude*.

7.1.16 Damping Depth

The *damping depth* may be defined as "the soil depth at which the temperature amplitude decreases to the fraction $1/e$ (i.e., $1/2.718 = 0.37$) of the temperature amplitude."

7.2 Soil Temperature Requirement of Field Crops

Soil temperature is an important abiotic factor that affects the entire physiology of plants starting from germination of seeds through root and shoot growth to plant yield. Soil temperature has a significant effect on processes like water absorption, mineralization, nutrient availability, and soil aeration. Each plant has specific temperature requirement at different growth stages for healthy growth and yield. Temperature requirement of crop plants varies with the plant species, cultivars, soil properties, and climatic conditions. The growth of almost all crop plants is drastically reduced at soil temperatures <9 °C and >50 °C. Alli and Omofunmi (2021) have published a review paper relating soil temperature to climate change, soil physical properties, and plant growth.

Based on the crop response, three soil temperature ranges may be defined:

(i) *Optimum range*: the temperature range under which plants thrive and produce best; the optimum temperature range for most of the crop plants is 20–30 °C.
(ii) *Growth range*: the temperature range under which plants can grow, but the plant growth and yield may not be optimum; it is broader than the optimum range (and includes the optimum range).
(iii) *Survival limit*: the minimum and maximum temperatures that can be reached without killing the plants, although the plants may not grow and produce anything.

The soil temperature ranges for some crops are given in Table 7.1.

Like plant growth and yield, soil temperature for seed germination may also be the minimum soil temperature at which seeds can germinate without delay, which is called *minimum soil temperature*. Below the minimum soil temperature, seeds fail to germinate. The minimum soil temperature varies with crop species.

Growth of peas, radish, and spinach is significantly reduced at soil temperatures <10 °C, while growth of beans, cucumber, eggplant, pepper, and watermelon is reduced at soil temperatures <16 °C. Minimum temperature required for attaining full maturity of corn is 11–13 °C, of cotton is 12–17 °C, of potato and oats is 5–6 °C, and of wheat and pea is 2-4 °C. For survival of most of the vegetable crops, the upper limit of soil temperature is 30–40 °C. At soil temperatures >45 °C, practically no vegetable crop can survive for more than 24 h. The optimum temperature ranges for the emergence, tillering, anthesis, and ripening of rice are 25–30 °C, 25–31 °C, 30–33 °C, and 20–25 °C, respectively. Table 7.2 shows soil temperatures for seeding and germination of different vegetable crops.

Table 7.1 The approximate soil temperature (°C) ranges for some crops

Temperature range	Corn	Rice and sorghum	Winter wheat	Cereal rye
Optimum range	25–35	20–30	15–27	12–18
Growth range	10–39	10–40	5–35	5–30
Survival range	0–43	0–43	−20 to 43	−20 to 40

Table 7.2 Soil temperature requirement of different vegetable crops for their seeding and germination

Crop	Soil temperature at seeding (°C)	Seeds germinate at (°C)	Optimum seed germination occurs at (°C)
Beans	27	15	32
Carrot	27	5	32
Cucumber	18	15-38	32
Lettuce	24	5	24
Okra	21	15–38	32
Onion	15	10–38	27
Pepper	18	15–32	27
Pumpkin	18	15–38	32
Tomato	15	10–38	27
Watermelon	18	15–43	32

Source: Adapted from Alli and Omofunmi (2021)

Table 7.3 Minimum soil temperature for planting of some crops

Agronomic crops	Minimum soil temperature (°C)	Horticultural crops	Minimum soil temperature (°C)
Spring wheat	2.8	Spinach	3.3
Spring barley	4.4	Radish	4.4
Rye	5.0	Lettuce	5.0
Oats	6.1	Onion	5.6
Alfalfa	7.2	Peas	5.6
Spring canon	10.0	Potato	7.2
Sugar beet	10.0	Cabbage	7.2
Field corn	12.8	Carrot	7.8
Soybeans	15.0	Sweet corn	12.8
Sunflower	15.6	Pepper	13.9
Millet	15.6	Snap beans	13.9
Sorghum	18.3	Tomato	13.9
Dry beans	21.1	Cucumber	14.4
		Pumpkin	15.6

Source: Adapted from Alli and Omofunmi (2021)

The minimum soil temperatures for crop germination for some agronomic and horticultural crops are presented in Table 7.3.

7.3 Soil Temperature and Microbial Activities

Soil temperature affects soil biota (microscopic and macroscopic) directly through its effect on physiological and enzymatic activities and indirectly through its effect on physic-chemical properties of soil, such as diffusion and solubility of nutrients, mineral weathering, and rate of evaporation. The population, composition, growth, and activity of several beneficial soil microorganisms respond strongly to changes in soil temperature. Soil temperature also impacts other factors like soil moisture and quantity and quality of organic matter, which further impacts microbial activities.

Various chemical and biochemical reactions and mineralization processes, which have significant bearing on the release of nutrient elements in soils, are modulated by soil temperature. Microbial activity generally increases with rise in soil temperature, but the temperature effect is modified by other factors like soil moisture and climatic conditions. For example, higher temperatures may not lead to higher decomposition rates of organic matter in dry soils. Freeze–thaw cycles in cold climates may result in the death of sensitive soil microbes, release of nutrient elements in nutrient/plant available form, increased organic matter decomposition, and N mineralization rates.

Each organism has its own optimum temperature range for its best biological functional performance, and beyond this temperature range, the organism becomes inactive. Low temperatures lead to dormancy or hibernation, and high temperatures cause irreversible cellular changes subsequently leading to the death of the organism (Schipper et al. 2014).

The optimum soil temperature range for microbial growth and activities is 15–30 °C. Microbial activity is severely restricted below 10 °C and becomes practically zero below 5 °C. This temperature (i.e., <5 °C) is sometimes called as *biological zero temperature*. At temperatures above 40 °C, several microbes become inactive. Majority of soil microorganisms cannot survive at temperatures below 0 °C and above 50 °C. *Soil solarization* (raising and maintaining soil temperatures above 50 °C for a few days) is an effective technique to control pests and fungal diseases of vegetables and fruit crops.

7.4 Soil Temperature and Plant Nutrition

Soil temperature has a strong influence on the source of nutrients in soil and their availability to plants. There are possibly three sources of nutrients in soil:

- The weathering of mineral component of soil
- The decomposition and mineralization of organic matter
- The externally applied nutrients as fertilizers or amendments

The primary source of all nutrients in soil, except C, H, O, and N, is the mineral component of soil. The C, H, and O are derived from atmospheric CO_2 and H_2O, and N is derived through N fixation. Most of the nutrient elements in the soil are held in a crystalline form within primary and secondary minerals. They are slowly released in soil solution through weathering processes, which in turn are strongly influenced by soil thermal regimes. For example, maximum release of K from the non-exchangeable form occurs above 30 °C.

The metabolic activities of microorganisms play an important role in the cycling of nutrients (e.g., N, S, P, B) in the soil through microbial decomposition of soil organic matter (SOM) and mineralization processes. They also ensure that the nutrients in soil are present in forms available to plants. The microbial activities as well as composition of soil organic matter are temperature controlled. The microbial decomposition of SOM is highly restricted at temperatures below 10 °C and stops practically below 5 °C. The increase in soil temperature above 10 °C increases metabolic activities of microorganisms and stimulates the availability of nutrients for plants. The process continues until a critically higher temperature is reached at which microbes become inactive. Studies have shown that with every 1 °C rise in soil temperature, there is around 10% loss of soil organic carbon and nitrogen. Bottner et al. (1995) concluded that a 3 °C increase in soil temperature would deplete the SOC pool in the surface soil by 28, 20, and 15% in humid, subhumid, and arid zones, respectively. Nevertheless, the dependence of SOM decomposition on soil temperature is highly variable due to the heterogeneous nature of SOM and diverse physicochemical properties of soil minerals. The nature of clay also has a significant effect on the relationship between SOM decomposition and soil temperature. Singh et al. (2017) reported that under a given set of soil moisture and temperature conditions, the SOM decomposition rate followed the following trend: smectitic soil > kaolinitic-illitic soil > allophanic soils. Decomposition was negatively correlated with the total sesquioxide (Fe and Al oxides) content of these soils.

Nitrification (conversion of NH_4^+ to NO_3^- by microbial oxidation) is highest at around 30 °C. The N-fixing bacteria are most active in warm (20–25 °C) and well-drained soils.

The dissolution of externally applied nutrients in soil solution is also temperature dependent. The solubility of nutrient elements in soil solution increases with the rise in soil temperature, and hence their availability to plants. Low temperatures not only retard K release but also decrease K uptake by the plant roots. At low soil temperatures, there is reduction in nutrient uptake by plants due to high soil water viscosity and low activity of root nutrient transport (Grossnickle 2000; Lahti et al. 2002).

7.5 Soil Temperature, Water Absorption, and Transpiration

The water absorption by plant roots depends on two processes: (a) water transport to plant roots and (b) ability of plant roots to absorb water. The hydraulic properties of soil are temperature dependent, mediated through surface tension-viscous flow mechanisms. It was known long ago since Sachs's classical experiments (Sachs 1875) cited by Kramer (1940) that low temperatures decreased water absorption by plants. Sachs (1875) made an interesting observation that plants from warmer climates showed wilting when exposed to 3–5 °C soil temperatures, while plants from cooler climates did not.

There are possibly two types of constraints responsible for reduced water uptake at low soil temperatures: *physical constraints* and *physiological constraints*. The *physical constraints* are associated with increase in water viscosity, surface tension, and density at low soil temperatures. Low soil temperatures may also modify soil particles, soil porosity, and interactive surface between liquid and solid, especially in soils with high clay contents, through shrinkage of soil matrix (Gao and Shao 2015). These low temperature-induced physical changes impact water fluxes through soils. The *physiological constraints* include increased viscosity of protoplasm, making it less permeable to water (Maximov 1929). The root membranes are less water permeable at low temperatures. It slows down the movement of water into and through the root cells. It is not clear, however, which of the two constraints, physical or physiological, dominates the other in reducing water uptake at low soil temperatures. Water absorption by plant roots increases with increase in soil temperature until about 35 °C. Above 35 °C, water absorption by most plant species is greatly reduced.

Transpiration rate in most of the plant species is significantly reduced at soil temperatures below 10 °C due to reduced water uptake (Kramer 1969).

7.6 Soil Temperature and Plant Growth

Soil temperature has a significant influence on plant growth and development. It influences the entire physiology of plants starting from germination, emergence and development of seedlings, and root and shoot growth to plant yield. Soil temperature also affects aging of plants. The temperature effect is more critical during early growth stages, viz. seed germination, seedling emergence, and early development of seedlings. There are several pathways, direct and indirect, by which soil temperatures may influence plant metabolism and growth. Soil temperature affects root growth, water and nutrient uptake by roots, photosynthesis, and transpiration. All these processes are reduced at low soil temperatures. All plants have minimum, optimum, and maximum temperatures for germination, plant growth, and yield. The growth of almost all crop plants is slowed down at soil temperatures <9 °C and >50 °C. The effect of soil temperature on selected plant growth stages is briefly discussed below:

7.6.1 Seed Germination

Seed germination is affected more by soil temperature than air temperature, while the seedling and plant growth is affected more by air temperature. All seeds require specific soil temperature for their optimum germination, below and above which seed germination is adversely affected. Depending on the response of seed germination, the soil temperature may be classified into three categories: minimum, maximum, and optimum. The *minimum* temperature for seed germination is the

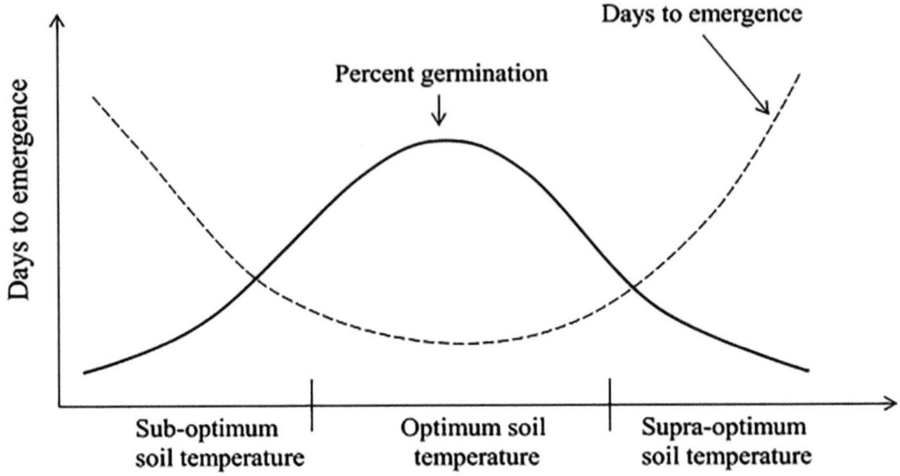

Fig. 7.1 Effect of soil temperature on seed germination

lowest temperature at which seeds germinate, but the time duration of germination is much longer than the optimum temperature. The *maximum* temperature is the highest temperature at which seeds fail to germinate. High temperatures may cause seed dormancy or even death of seeds. Even if the seeds germinate at high soil temperatures, the seedlings may die fast from the heat stress. The *optimum soil* temperature for seed germination is one at which germination is 100%. It is an ideal temperature. But the *realistic*, i.e., *practical*, soil temperature is one at which seed germination is 70–80%. Seeds sown under sub- or supraoptimal soil temperature conditions show delayed germination, low germination percentage, and slow growth (Fig. 7.1) maybe due to seed desiccation, fungal infection, pest attack, or suppression by weeds (if weeds grow faster than seedlings).

The optimum temperature for seed germination and plant growth varies with the crop species, soil types, and environmental conditions. Temperate crops germinate between 0 and 35 °C, while tropical crops germinate between 10 and 45 °C. The germination of temperate and tropical seeds is drastically reduced at soil temperatures above 35 and 45 °C, respectively.

The minimum temperature for germination of seeds of wheat, barley, peas, lentil, mustard, and clovers is 0–3 °C; for potato, sunflower, and sugar beet is 5–8 °C; for corn, millets, and beans is 10 °C; and for rice, cotton, sorghum, and sesame is 11–14 °C.

The optimum temperature range for germination of seeds of wheat and peas is 14–20 °C; for corn and millets is 26–34 °C; for potato is 16–21 °C; for tomato, radish, carrot, onion, mustard, lettuce, asparagus, and celery is 18–25 °C; and for rice, cotton, sugarcane, sorghum, and melon is 25–33 °C.

Minimum and optimum soil temperatures for germination of some seeds are given in Table 7.4.

Table 7.4 Minimum and optimum soil temperatures (°C) for germination of some crops

Crop	Minimum temperature	Optimum temperature
Cereals and oilseeds		
Wheat	4	12–25
Barley	3–5	12–25
Corn	16	30–35
Oats	5	20–24
Pearl millets	10–12	30–35
Soybean	10–15	25–30
Mustard/canola	10	15–20
Sunflower	7–10	20–30
Forage crops		
Alfalfa	1–8	25
Red clover	3–7	25
Sweet clover	1	18–25
White clovers	5	18–20
Fescues	3	13–18
Orchard grass	4	18–20
Vegetable crops		
Beans	8–10	16–30
Beet	4	10–30
Cabbage	4	7–35
Carrot	4	7–30
Cauliflower	4	7–30
Celery	4	15–21
Lettuce	2	4–27
Potato	7	18
Tomato	10	21–27
Cucumber	16	25–35
Eggplant	16	24–32
Lady's finger (okra)	24	30–35

(Compiled from different sources)

7.6.2 Root and Shoot Growth

Soil temperature has a significant effect on root growth. Root morphology (i.e., root length, root mass, and branching) and root functions (water and nutrient absorption) and metabolism (enzymatic activity) are significantly affected (McMichael and Burke 1998). The branch roots are more sensitive to soil temperature than the primary roots. There is a certain minimum temperature below which no root growth occurs. Above this minimum temperature, root growth increases almost linearly with increasing soil temperature until an optimum temperature is reached above which root growth starts decreasing and ultimately ceases completely. Roots growing at low soil temperatures are short, stubby, and less branched and may suffer death of root cortex and wilting of plant (Christiansen 1963). The initiation of adventitious roots is also influenced by soil temperature. For example, the optimum temperature range for the growth of adventitious roots in corn is 10–30 °C. Reduction in root growth at low soil temperatures may result from low nutrient concentrations in root cells and reduced cell multiplication. At soil temperatures higher than the optimum, the root elongation rates are reduced but root branching may be enhanced (Nielsen 1974). The optimum temperature range varies with the plant species and genotype (Cooper 1973; Glinski and Lipiec 1990). For example, optimum soil temperature ranges for root growth of cotton and sunflower are 28–35 °C and 23–25 °C, respectively (McMichael and Burke 1998). Optimum root growth occurs at approximately 35 °C for subtropical plants, 27.7 °C for warm temperate plants, and 20 °C for cool temperate plants. The impact of soil temperature on root growth varies with the available soil water content. Strong interactions among soil temperature, available soil water content, pathogen activity, and root growth have been observed. Glinski and Lipiec (1990) observed that at 30 °C, corn root growth increased with increasing plant available water content (PAWC), but no such effect was observed at 15 °C.

The effect of soil temperature on root growth also varies with the type of root system. Low temperatures encourage white succulent roots having little branching, while high temperatures encourage a browner, finer, and much more freely branching root systems.

Soil temperature also affects shoot growth. Effect of soil temperature on root/shoot ratio varies with the air temperature. There is a strong interaction among soil temperature, air temperature, and root and shoot growth. Studies indicate that the optimum soil temperature for shoot growth is a function of the root/shoot ratio. Increasing soil temperature decreases the root/shoot ratio, while increasing air temperature increases the root/shoot ratio.

7.6.3 Photosynthesis

Photosynthesis in plants is strongly influenced by soil thermal regimes. The extreme soil temperatures (very low and very high) significantly reduce photosynthesis. Reduction in photosynthesis at low soil temperatures is associated with the stomatal closure due to reduced water absorption by plant roots. Plant cells fail to maintain turgidity optimum to keep the stomata open. Low temperatures also elevate abscisic acid concentrations in root and shoot tissues, which in turn reduces stomatal conductance (Starr et al. 2004). It leads to reduced rate of CO_2 absorption by leaves and reduced rate of photosynthesis. At high soil temperatures, on the other hand, photosynthesis reduces due to the decrease in leaf chlorophyll and carotenoid contents and increase in leaf soluble proteins (Ruter and Ingram 1992). The impact of soil temperature on photosynthesis is reflected in plant growth and yield.

7.7 Factors Affecting Soil Temperature

Soil temperature is a dynamic soil physical property and is influenced by a number of environmental and soil factors including soil color, slope characteristics, soil depth, state of looseness/compaction, moisture, vegetation cover, solar radiation, etc. Use of organic manures and fertilizers along with soil management practices also impacts soil temperature. Further, the impact of any given factor on soil temperature is strongly modified by other factors because of the complex interactions occurring among various temperature-influencing factors. Hence, it becomes difficult to isolate the impact of any individual factor on soil temperature. Knowledge of various factors affecting soil temperature is important for the management of soil thermal regimes. Understanding of the mechanisms and interactions among different factors and soil thermal regimes is needed for deciding effective soil temperature management strategies. These data are also required in developing yield-forecasting models.

7.7.1 Environmental Factors

The solar radiation is the primary source of thermal energy in soil. The solar radiation received on the soil surface depends on factors like latitude, longitude, slope characteristics, weather parameters, and duration of exposure. The higher the solar energy received, the higher the soil temperature. The impact of different factors on the amount of solar radiation received on the soil surface is briefly described below:

7.7.1.1 The Angle of Incidence of Solar Radiation

Solar radiation received on the soil surface depends on the angle at which solar radiation meets the soil surface. The angle of incidence of solar radiation determines the land area exposed to radiation. The higher the angle of incidence, the smaller the exposed area and the more the solar radiation received at earth's surface per unit area. The area covered by incident radiation at 45° angle is about 40% more than by the incident radiation at 90° angle. Therefore, when solar radiation is perpendicular, earth's surface receives more radiation per unit area and is warmer than when solar radiation reaches earth's surface at acute angles. The angle of incidence of solar radiation depends on the latitude and altitude of location, sun's position with respect to earth's surface, and steepness and direction of slope. Duration of exposure of soil surface to solar radiation is also a determining factor. Some estimates suggest that in temperate regions, soil receives around 100–800 langleys/day solar radiation.

7.7.1.2 Direction of Slope

Southern (i.e., south-facing) slopes receive more solar radiation than the northern (i.e., north-facing) slopes. Therefore, the soil temperatures of southern slopes are higher than the northern slopes. It can be easily observed by snow covers, which stay much longer on northern slopes than on southern slopes in Himalayas. Northern slopes, therefore, remain moist for a longer period of time and support thicker vegetation and forests than the southern slopes, which become dry very fast due to longer exposure to the sun.

7.7.1.3 Insulation Effect

The air, water vapors, clouds, dust, snow, etc. in the earth's atmosphere impact solar radiation received on the soil surface through insulation effect. Insulated soils show diurnal or seasonal temperature fluctuations to a lesser degree than non-insulated soils. Insulated soils have relatively lower temperature during summers and higher temperature during winters than non-insulated soils.

Water vapors, fog, clouds, dust, and different gas molecules in the atmosphere reflect, absorb, and/or scatter the incoming solar radiation, thereby decreasing the net radiation reaching the ground. They obstruct the long-wave radiation from escaping the soil surface and keep them near the soil surface. That is why cloudy days are cooler and cloudy nights are warmer than the clear days and nights, respectively. The presence of large water bodies in the area keeps the surrounding air more saturated with water vapors.

7.7.1.4 Evaporation and Condensation

Evaporation is an endothermic process and so is the melting of snow. Both the processes show a cooling effect in soil. Condensation of water vapors and freezing of water, on the other hand, are exothermic processes. They raise soil temperature by as high as 5 °C or even more.

7.7.2 Soil Factors

7.7.2.1 Soil Moisture

Soil moisture regulates soil temperature through its effect on thermal properties, including heat capacity, thermal conductivity, thermal diffusivity, and thermal retentivity as well as evaporation and condensation processes in soils. Thermal conductivity, thermal diffusivity, thermal retentivity, and heat capacity of soils increase with soil water content. Increase in thermal conductivity and heat capacity

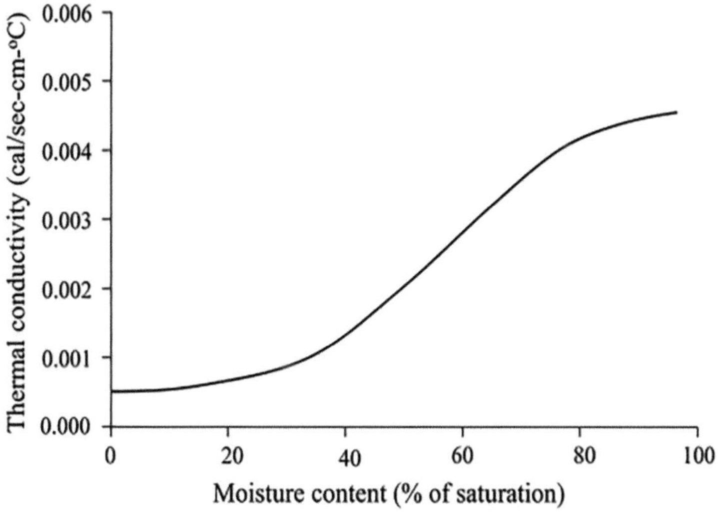

Fig. 7.2 Effect of soil moisture content on the thermal conductivity of a medium-textured soil

is more than increase in thermal diffusivity and thermal retentivity. Water molecules increase thermal conductivity by making bridges between soil particles. Thermal conductivity of soils at −10 bar matric potential is almost equal to that of water. Relationship between thermal conductivity and soil moisture content is shown in Fig. 7.2. Thermal diffusivity is maximum at about −1.0 bar matric potential and decreases as matric potential falls below −1.0 bar.

The major effects of soil moisture on soil thermal regime are summarized below:

- A moist to wet soil warms up much slowly than a dry soil because of its higher heat capacity, thermal conductivity, and thermal retentivity. A moist soil may conduct heat 6–8 times faster than a dry soil. Solid rock conducts heat much faster than a wet soil.
- Diurnal temperature fluctuations at soil surface are comparatively low in moist/wet than in dry soils.
- Evaporation at the surface of wet soils keeps soil temperature 3–6 °C lower than the dry soil. Evaporation is an endothermic process.
- The wet soils, because of high specific heat of water, are cooler in summers and warmer in winters compared to dry soils. Wet soils exhibit more uniform temperature profiles with depth than dry soils.

7.7.2.2 Soil Texture and Structure

Soil texture and soil structure also influence soil thermal regimes but to a lesser extent than soil water content. The effects are summarized below:

- Mineral soils have higher thermal conductivity than organic soils. Mineral soils, therefore, exhibit higher diurnal temperature fluctuations on the soil surface.
- Clay particles usually have higher heat capacity than sand particles at given moisture content and bulk density. Sandy soils being less porous than clayey soils have higher thermal conductivity. The sand, clay, and moisture content impacts soil temperature in a complex way.
- Soil compaction increases thermal conductivity by increasing physical contact among soil solids, and the soil solids conduct more heat than air. Temperature fluctuations are, therefore, more in compact than loose soils.
- Platy and blocky structure shows greater thermal conductivity than the porous granular structure, because of better physical contact among soil solids. Diurnal temperature fluctuations are, therefore, more in soils having platy and blocky structure.

Approximate average data for thermal properties of some soil components are shown in Table 7.5.

7.7.2.3 Soil Color

A part of solar radiation received on the soil surface is absorbed by soil and a part is reflected back into the atmosphere. The absorption or reflection of solar radiation depends on the color of the soil surface. Bright colors are more reflective of radiation than dark colors. Dark soils, such as peat and muck soils or mineral soils rich in organic matter, reflect less and absorb more solar radiation and are, therefore, warmer than bright- or light-colored soils. Dark soils also warm up fast.

Table 7.5 Approximate average data of thermal properties of some soil components

		Heat capacity				
Soil component	Density (g/cc)	Specific heat capacity (cal/g-°C)	Volumetric heat capacity (cal/cm^3-°C)	Thermal conductivity (cal/cm-sec-°C)	Thermal diffusivity (cm^2/s)	Thermal retentivity (s/cm^2)
Soil-forming minerals	2.65	0.20	0.53	0.010	0.019	52
Water	1.00	1.00	1.00	0.0014	0.0014	713
Ice	0.917	0.505	0.474	0.005	0.0105	95
Dry air	0.0012	0.25	0.0003	0.00005	0.167	6

Source: Adapted from Kohnke (1968)

7.7.2.4 Biological Activity in Soils

Biological activities generate heat in soil. Biologically active soils (i.e., well-aerated soils having adequate moisture and rich in organic matter and plant nutrients), therefore, exhibit higher temperature under a given set of conditions than biologically inactive/less active soils. Applications of composts and manures keep soils warmer.

7.7.2.5 Soil Cover

Bare lands warm up faster than covered lands. The soil covers, such as vegetation and mulch, decrease net solar radiation reaching the soil surface.

The *vegetation cover* impacts soil temperature through reflection of incident radiation, decreasing the depth of penetration of solar radiation through the canopy, increasing latent heat transfers through evapotranspiration, and decreasing heat loss from soil through insulating effect. Thus, vegetation cover lowers diurnal temperature fluctuations and keeps soil warmer during winters and cooler during summers as compared to dry soils.

Mulch materials spread on the soil surface decrease both the positive and negative heat fluxes in soils. Mulches reflect a part of solar radiation, thereby decreasing radiation reaching the soil surface (less warming). They trap the solar radiation escaping the soil surface in the form of long waves (warming effect). They also reduce evaporation from the soil surface (less cooling effect). The net soil temperature under mulch, therefore, is the resultant of balance among these processes. The soil temperature under mulch may be higher or lower than in a bare or cropped soil without mulch depending on the quality and quantity of mulch material. Soil temperature under transparent plastic sheet mulch is usually higher than under straw mulch. The impact of plastic sheet mulch on soil temperature varies with the color, thickness, and mode of placement of sheet on the soil surface.

The mulched soils generally experience low diurnal temperature variations and remain cooler during the day and warmer during the night as well as cooler during summers and warmer during winter months. They avoid temperature extremities compared to un-mulched bare soils.

The ability of mulch material to reflect the incident radiation is determined by its "albedo." The *albedo* refers to "the ratio of reflected to incident radiation." Higher albedo indicates higher reflection of incident radiation and, thus, a cooler soil, and vice versa. Light-colored and dry mulches are more reflective than dark-colored and moist mulches. Albedo values of some materials are shown in Table 7.6.

Table 7.6 Albedo values of some natural materials

Material	Albedo	Material	Albedo
Fresh snow	0.75–0.95	Water	0.03–0.01
Bare soil	0.12–0.25	Stubble field	0.15–0.17
Dry soil	0.20–0.24	Wet soil	0.12–0.14
Dry grass	0.16–0.19	Green grass	0.16–0.27
Grain crops	0.10–0.25	Forests	0.05–0.20

Source: Adapted from Baver et al. (1976)

7.7.2.6 Soluble Salts

Soluble salts may influence soil thermal regimes indirectly through their effect on processes like evaporation and condensation, and belowground biological activities. Higher salt concentration lowers evaporation, soil fertility, and biological activities in soils. Such changes in soil may have a significant impact on soil temperature.

7.8 Management of Soil Temperature

The management of soil temperature aims at optimizing soil thermal regimes for the normal growth and performance of plants. The energy balance equation suggests that soil temperature is the resultant of incoming (solar radiation) and outgoing radiation (long-wave radiation on the soil surface). If the incoming solar radiation is more than the outgoing long-wave radiation plus the reflected radiation on the soil surface, energy balance is positive and soil temperature rises. Vice versa is also true. Soil temperature, thus, may be regulated by

(i) Modifying heat exchange processes between soil and atmosphere, i.e., net incident solar radiation on the soil surface and outgoing long-wave radiation from soil to the atmosphere
(ii) Changing the soil thermal properties which impact heat fluxes and heat retention within the soil profile

These changes may be easily achieved by the commonly practiced field operations like tillage, compaction, mulching, irrigation, drainage, and use of cover crops.

The choice of management options depends on whether the temperature of soil needs to be increased or decreased. It may, however, be kept in mind that crop behavior under field conditions cannot be related to the average soil temperatures. The alternating soil temperatures (day and night, morning and evening temperatures), especially the temperature peaks, have quite a different effect on crop growth aspects than the constant or average soil temperatures as are usually studied under controlled conditions. Such results should be used with caution when applied to field conditions.

7.8.1 Tillage

Tillage practices (plowing, discing, harrowing, etc.) loosen the soil, decrease bulk density, and increase soil porosity, thereby bringing significant changes in soil thermal properties. Loose soils, as discussed earlier, possess low thermal conductivity and heat capacity. They decrease heat flux into the soil and increase diurnal temperature fluctuations on the soil surface. Consequently, the maximum and minimum temperatures on the surface of a tilled soil are higher than a soil without tillage, i.e., zero-till soil. Compact soils have high thermal conductivity and heat capacity, and thus high heat flux into the soil. Compact soil surface, as in the case of zero-till soils, therefore, has higher soil temperature at night and lower temperature in the daytime compared to a well-pulverized loose soil. Effect of tillage on soil temperature varies with the season. Fall tillage may result in lower maximum and higher minimum soil temperatures than spring tillage (Wall and Stobbe 1984). Temperature rises fast during spring season in a fall-tilled soil than in an untilled soil. The temperature difference at 10 cm depth may be as much as 5 °F.

7.8.2 Land Shaping

A ridged soil surface has higher temperature than a flat surface field. The ridge-furrow system in general increases soil temperature by increasing the area exposed to radiation and decreasing soil moisture. Direction and degree of slope have a significant effect on soil temperature. Soil temperature varies significantly between ridge and furrow depending on the row direction and time of the day. Soil temperature is higher at the ridge than in the furrow. The south- and west-facing slopes have higher soil temperature than the north- and east-facing slopes.

7.8.3 Mulching and Vegetation Cover

Mulches and vegetation regulate soil temperature through

(i) Interception of incident radiation
(ii) Modification in albedo
(iii) Reduction in latent heat transfer by evaporation
(iv) Conservation of soil moisture

Mulches and vegetation cover intercept incident solar radiation on the soil surface. Mulches having high albedo (e.g., straw mulch) also reflect incoming solar radiation. Resultantly, the net flux of incident solar radiation on the soil surface is reduced giving a cooling effect. Mulches also reduce outgoing long-wave radiation. Mulches reduce evaporation and conserve soil moisture. Wet soils are cooler than dry soils. The nature and color of mulch material have a significant effect on soil thermal regime. The transparent plastic mulch produces a glasshouse effect by allowing solar radiation to penetrate it and trapping outgoing long-wave radiation. It keeps soil temperature higher than under the black plastic and straw mulch, and in a bare soil. The temperature differences at 5 cm soil depth may be as high as 4–10 °C. Cover crops keep surface soils cool through shading and insulating effect.

7.8.4 Irrigation and Drainage

Irrigation modifies soil thermal regime by increasing the soil moisture content. The temperature of irrigation water also influences soil temperature. Irrigation with cold water during summers results in lowering of the soil temperature. Drainage removes water and modifies the air-water balance in the soil pores. Removal of excess water by surface or subsurface drainage increases soil aeration, which in turn warms the soil surface.

7.9 Evaluation of Soil Temperature

Soil temperature is a dynamic soil property which keeps on changing in response to the changing meteorological parameters such as solar radiation, clouds, and precipitation. It shows high spatiotemporal variability. Its precise and accurate evaluation is important for proper crop planning (e.g., selection of crops and their sowing times) and modeling crop responses to soil temperatures representing field conditions (Wanjura et al. 1970). Several technologies are available to measure soil temperature. Soil temperature is generally measured with *thermometers, thermocouples,* or *thermistors*. Other devices, such as *distributed temperature sensing* (DTS), *remote sensing*, and *satellite monitoring*, are also available for temperature measurements at a large scale. It is possible to generate soil temperature data at any point in time and space, making a series of observations at one point, inclusive of surface and air temperatures. The online platforms for assessing soil temperature data by the farmers in real time frame have made assessment of soil thermal regimes more effective and decision-making about the type of crop and time of sowing more accurate.

Some devices used to measure soil temperature are briefly discussed below:

7.9.1 Thermometers

Thermometers are nonelectrical devices to measure soil temperature. They function on the principle of thermal expansion of liquids or solids. They are simple, less expensive, easy to operate, and used for routine fieldworks where very high accuracy of temperature measurement is not required.

Thermometers are of two types: contact type or noncontact type. The contact-type thermometers again are of two types: liquid type and solid type. The *liquid-type* thermometers use liquids, such as mercury or alcohol, and measure soil temperature based on thermal expansion of these liquids. The *solid-type* thermometers use metal strips and work on the principle of expansion of solids depending on their coefficient of thermal expansion. In bimetallic thermometers, two rods or strips of metals having different coefficients of thermal expansion are firmly welded together. As the soil temperature changes, the two metals expand differentially, creating deformation in the bimetallic rod or strip, and this deformation is a measure of soil temperature.

7.9.2 Thermocouples

A *thermocouple* is a sensor to measure soil temperature. It works on the principle of thermoelectric *effect*, also called *Seebeck thermoelectric effect.* Thomas Johann Seebeck, a German Physician-turned physicist, in 1821 observed that "when any conductor is subjected to a thermal gradient, a voltage is generated," a process known as *thermoelectric effect.* The voltage so generated is a measure of temperature.

A thermocouple consists of two dissimilar metals (such as copper and constantan) that are joined together forming two junctions, viz. *measurement junction* and *reference junction.* When these two metal junctions are kept at two different temperatures, a voltage is generated due to *thermoelectric effect*, which can be easily measured with the help of a potentiometer or voltameter. This voltage is the measure of soil temperature. Different metals are used for measuring different temperature ranges.

7.9.3 Thermistors

Thermistors are resistors whose resistance is strongly temperature dependent. They are made of ceramic-like semiconducting materials, such as oxides of manganese, nickel, and cobalt. They operate on the principle that the electrical resistance of a semiconductor changes with the change in the temperature, and this change in resistance is a measure of temperature. In some cases, there is decrease in electrical resistance with the increase in temperature, while in others, electrical resistance increases with the temperature. The thermistors of the first kind are known as *negative temperature coefficient* (NTC) thermistors, while the second type is known as *positive temperature coefficient* (PTC) thermistors. The NTC thermistor was discovered by Michael Faraday in 1833. A commercially viable thermistor was invented by Samuel Ruben in 1930.

Thermistors are more sensitive than thermometers or thermocouples in measuring temperature. They are also very precise in measuring temperature within a temperature range of −90 °C to 130 °C.

7.9.4 Distributed Temperature Sensing (DTS) Systems

The DTS systems are optoelectronic devices which measure temperatures by means of optical fibers, which function as linear sensors. Temperatures are recorded along the optical sensor cable as a continuous profile and not at points. Thus, the DTS enables more detailed quantitative measurements of soil temperature at small to medium scales (1 m up to 30 km) compared to traditional point measurements of soil temperature. The DTS instruments are based on the principle of *Raman effect*. The optical fiber cables can be placed at shallow depths in the fields with the help of special plow and used to measure spatial distribution of soil temperature over long distances. The DTS systems can locate the temperature to a spatial resolution of 1 m with accuracy to within ±1 °C at a resolution of 0.01 °C. For details, one may refer to a review article published by He et al. (2018).

7.9.5 Remote Sensing

Remote sensing techniques can determine variation in soil thermal regimes over large areas. Remote sensing of reflected light from soil surface indicates thermal response of the topmost soil layers (a few molecular layers thick). The thermal infrared (TIR) light waves, on the other hand, provide information about the thermal variations extending to varying shallow depths below the ground surface. The TIR is affected by cloud cover. Microwave remote sensing from satellites has an advantage over TIR as it is not affected by the cloud cover. Such techniques are useful to make frequent temperature observations on large areas.

7.10 Time Lag

Heat wave takes time to travel from soil surface to deeper soil layers due to the relatively high heat capacity and low heat conductance of the soil. Resultantly, the occurrence of temperature peaks in soil gets delayed as compared to that in air. This time lag between the occurrence of temperature peak (maximum or minimum temperature) in air or on the soil surface and in soil at a given depth is known as the *time lag* or *time delay*. Mathematically,

$$\text{Time lag} = -z / d \tag{7.1}$$

where z is the soil depth and d is the damping depth (Sect. 7.1.16).

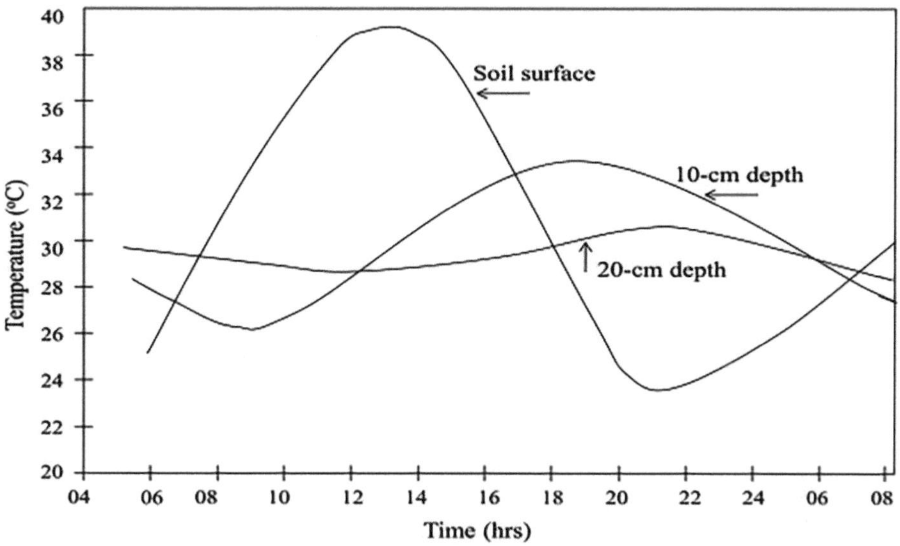

Fig. 7.3 Effect of soil depth on diurnal temperature amplitude and time lag in a puddled and flooded rice soil (adapted from Sharma and De Datta 1991)

In an arable land, the time lag between temperature peak in the air and on the soil surface is almost zero. The time lag increases with soil depth, as it takes time for the heat wave to move through the soil. In arable soils, rate of penetration of heat wave within the soil profile is estimated to be around 3.3 cm/h.

Time lag is more in a wet than in a dry soil. Sharma and De Datta (1991) observed a time lag of 1.2, 4.6, 6.4, 8.7, 10.5, 11.2, and 11.6 h, respectively, between maximum air temperature 50 cm above soil surface and at 0, 5, 10, 15, 20, 25, and 30 cm depth in a 5 cm flooded puddled rice soil (Fig. 7.3). Chaudhary and Sandhu (1982) observed time lag in a bare sandy loam soil of 2.0, 2.5, 3.0, and 6.0 h at 5, 10, 15, and 30 cm soil depth, respectively.

The occurrence of time lag between air and soil temperature peaks has a strong implication while measuring soil temperatures. The time of measurement of temperature peaks in soil at different depths must consider the time lag for that soil. Time lag is influenced by texture, structure, organic matter content, and water status of soil.

7.11 Question Bank

7.11.1 Short Questions:

(i) Define the following:

(a) Heat
(b) Temperature

(c) Calorie
(d) Thermal capacity
(e) Thermal diffusivity
(f) Radiation
(g) Solar constant
(h) Amplitude
(i) Damping depth

(ii) Write a short note on the soil temperature requirement of crop plants.
(iii) Briefly describe the impact of soil temperature on

(a) Soil microbial activities
(b) Plant nutrition
(c) Water absorption and transpiration by plants
(d) Seed germination
(e) Root and shoot growth of plants
(f) Photosynthesis

(iv) What is the role of environmental factors on soil thermal regime?
(v) What soil factors influence soil temperature and how?
(vi) What is albedo? What is its role in soil temperature? How will you modify soil albedo?
(vii) What do you understand by the management of soil temperature? How do tillage systems affect soil temperature?
(viii) What is the effect of the following factors on soil temperature:

(a) Mulching and vegetation cover
(b) Irrigation and drainage

(ix) How is soil temperature measured? Briefly explain the principle and working of

(a) Soil thermometers
(b) Thermocouples
(c) Thermistors
(d) DTS systems
(e) Remote sensing

(x) What do you understand by *time lag* in the occurrence of temperature peaks at different depths? What is the role of soil water content in temperature lag? What is the significance of time lag in soil temperature?

7.11.2 Briefly Explain Why?

(i) Soil temperature impacts the mineralization processes in soil.
(ii) Nutrient availability in soils is significantly influenced by soil temperature.

(iii) Water uptake is reduced at low soil temperatures.
(iv) Seeds sown under sub- or supraoptimal soil temperature conditions show delayed germination, low germination percentage, and slow growth.
(v) Mulches with high albedo keep surface soil relatively cool.
(vi) Ridge-furrow system in general keeps soil temperature relatively high when compared with temperature of flat soil surface.
(vii) Tillage practices which loosen soil increase diurnal temperature fluctuations on the soil surface.
(viii) Biologically active soils exhibit higher temperature under a given set of conditions than biologically inactive/less active soils.
(ix) Dark-colored soils show higher soil temperature than light-colored soils.
(x) Temperature fluctuations are more in compact than loose soils.
(xi) *Soil solarization* is an effective technique to control pests and fungal diseases of vegetables and fruit crops.
(xii) Heat wave takes time to travel from the soil surface to deeper soil layers, a process called temperature time lag.

7.11.3 Fill in the Blanks:

(i) .... refers to the kinetic energy of the molecules of a substance due to their random motion.
(ii) .... refers to the condition of a body which determines the transfer of heat to or from other bodies.
(iii) The amount of heat that is present in a body is known as ....
(iv) The energy emitted by a body above absolute zero temperature in the form of electromagnetic waves is called ....
(v) The range of wavelength of shortwave radiation is ....
(vi) The range of wavelength of long-wave radiation is ....
(vii) The soil depth at which the temperature amplitude decreases to the fraction $1/e$ (i.e., $1/2.718 = 0.37$) of the temperature amplitude is known as ....
(viii) The survival temperature range for rice is ....
(ix) The survival temperature range for wheat is ....
(x) The optimum temperature range for rice is ....
(xi) The optimum temperature range for corn is ....
(xii) The optimum temperature range for wheat is ....
(xiii) The optimum soil temperature range for microbial growth and activities is ....
(xiv) The soil temperature at which microbial activity becomes practically zero is called ....
(xv) With every 1 °C rise in soil temperature, there is around ......% loss of soil organic carbon and nitrogen.
(xvi) Low soil temperatures .... K release in soils as well as K uptake by crop plants.

(xvii) The minimum temperature for germination of seeds of wheat, barley, peas, lentil, mustard, and clovers is ....
(xviii) The minimum temperature for germination of seeds of rice, cotton, sugarcane, sorghum, and melon is ....
(xix) The branch roots are .... sensitive to soil temperature than the primary roots.
(xx) Increasing soil temperature .... the root/shoot ratio.
(xxi) Increasing air temperature ...... the root/shoot ratio.
(xxii) Low temperatures .... abscisic acid concentrations in root and shoot tissues.
(xxiii) High abscisic acid .... stomatal conductance.
(xxiv) Low stomatal conductance is responsible for ...... photosynthetic rates at low soil temperatures.
(xxv) High soil temperatures ...... photosynthesis due to reduction in leaf chlorophyll and carotenoid contents.
(xxvi) High albedo on the soil surface results in .... soil temperature.
(xxvii) A *thermocouple* is a sensor to measure soil temperature, which works on the principle of thermoelectric *effect,* also called ....
(xxviii) The time lag in the occurrence of temperature peaks on the soil surface and deeper soil layers is ...... in dry than in wet/flooded soils.

[**Key:** i. Heat, ii. Temperature, iii. thermal capacity or heat capacity, iv. radiation, v. 0.3–4 μ, vi. 6.8–100 μ, vii. damping depth, viii. 0–43 °C, ix. −20 to 43 °C, x. 20–30 °C, xi. 25–35 °C, xii. 15–27 °C, xiii. 15–30 °C, xiv. biological zero temperature, xv. 10, xvi. decrease, xvii. 0–3 °C, xviii. 25–33 °C, xix. more, xx. decreases, xxi. increases, xxii. elevate/increase, xxiii. reduces, xxiv. low, xxv. decreases, xxvi. lower, xxvii. Seebeck effect, xxviii. more]

7.11.4 State Whether the Following Statements Are True (T) or False (F):

(i) The root membranes are less water permeable at low temperatures.
(ii) The growth of almost all crop plants is slowed down at soil temperatures <9 °C and >50 °C.
(iii) Seed germination is affected more by soil temperature than air temperature, while the seedling and plant growth is affected more by air temperature.
(iv) The *realistic optimum soil* temperature for seed germination is one at which germination is 100%.
(v) The seeds fail to germinate at soil temperatures below the optimum temperature range.
(vi) *Survival limit* is the soil temperature range under which plants can grow, but the plant growth and yield are below normal.

(vii) *Growth range* is the soil temperature range under which plants thrive and produce best.
(viii) At soil temperatures >45 °C, practically no vegetable crop can survive for more than 24 h.
(ix) The optimum soil temperature range for microbial growth and activities is 5–45 °C.
(x) Microbial activity becomes practically zero below 5 °C.
(xi) Heat and temperature are synonymous terms.
(xii) The thermal capacity, heat capacity, and specific heat capacity are synonymous terms.
(xiii) Radiation is the energy emitted by a body above absolute zero temperature in the form of electromagnetic waves.
(xiv) The temperature difference between the average and peak (maximum or minimum) temperature on a daily basis is called diurnal fluctuation.
(xv) Soil temperature is an important *abiotic* factor that affects the entire physiology of plants.
(xvi) Wheat plants can survive at soil temperature as low as −20 °C, while rice plants cannot survive at soil temperatures <0 °C.
(xvii) The soil temperature at which microbial activity becomes practically zero is called biological zero temperature.
(xviii) S*oil solarization* is an effective technique to control pests and fungal diseases of vegetables and fruit crops.
(xix) With every 10 °C rise in soil temperature, there is around 10% loss of soil organic carbon and nitrogen.
(xx) The N-fixing bacteria are most active in warm (20–25 °C) and well-drained soils.
(xxi) Low soil temperatures retard K release but normally do not affect K uptake by the plant roots.
(xxii) Crop plants from warmer climates may show wilting when exposed to low (3–5 °C) soil temperatures while plants from cooler climates may not.
(xxiii) Transpiration rate in most of the plant species is significantly reduced at soil temperatures below 10 °C due to reduced water uptake.
(xxiv) Increasing soil temperature increases the root/shoot ratio, while increasing air temperature decreases the root/shoot ratio.
(xxv) Low temperatures elevate abscisic acid concentrations in root and shoot tissues.
(xxvi) Low abscisic acid reduces stomatal conductance.
(xxvii) Low stomatal conductance is responsible for low photosynthetic rates at supranormal soil temperatures.
(xxviii) There is reduction in photosynthetic rates at high soil temperatures due to the reduction in leaf chlorophyll and carotenoid contents.
(xxix) The time lag in the occurrence of temperature peaks on the soil surface and deeper soil layers is more in dry than in wet/flooded soils.

[**Key:** i. T, ii. T, iii. T, iv. F, v. F, vi. F, vii. F, viii. T, ix. F, x. T, xi. F, xii. F, xiii. T, xiv. X, vi. F, xv. T, xvi. T, xvii. T, xviii. T, xix. F, xx. T, xxi. F, xxii. T, xxiii. T, xxiv. F, xxv. T, xxvi. T, xxvii. F, xxviii. T, xxix. F]

7.11.5 Multiple-Choice Questions:

(i) Temperature of a body refers to

(a) The kinetic energy of the molecules of a substance due to their random motion
(b) The condition of a body which determines the transfer of heat to or from other bodies
(c) The amount of heat that is present in a body
(d) The amount of heat required to raise the kinetic energy of molecules

(ii) Around 99% solar radiation is received on the soil surface in

(a) Shortwave radiation
(b) Long-wave radiation
(c) Combination of short- and long-wave radiation
(d) Infrared radiation

(iii) Temperature amplitude refers to

(a) Temperature fluctuation in soil on a daily basis
(b) Average temperature fluctuation in soil on a seasonal basis
(c) Temperature difference between the average and peak temperature
(d) Difference in maximum and minimum daily temperature

(iv) *Growth range* refers to the temperature range

(a) Under which plants thrive and produce best
(b) Under which plants survive but may not produce any yield
(c) Under which plants grow but the plant growth and yield may not be optimum
(d) Below and above which plants fail to survive

(v) *Survival limit* refers to the soil temperature

(a) Below which plant growth and yield are retarded
(b) At which plants wilt and die
(c) Above which plant growth and yield are normal
(d) Minimum or maximum temperature below or above which plants die

(vi) The soil temperature range optimum for plant growth is

(a) Broader than the growth range
(b) Narrower than the growth range

(c) Broader than the survival range
(d) Same as the growth range

(vii) Albedo of fresh snow is

(a) Equal to albedo of water
(b) Higher than albedo of water
(c) Lower than albedo of water
(d) Higher or lower than albedo of water depending on the quality of water

(viii) Outgoing radiation from soil is generally in the form of

(a) Ultraviolet radiation
(b) Shortwave radiation
(c) Long-wave radiation
(d) Combination of short- and long-wave radiation

(ix) Thermal conductivity of soil increases with the

(a) Increase in total porosity of soil
(b) Decrease in bulk density of soil
(c) Decrease in soil moisture content
(d) Increase in bulk density of soil

(x) The *time lag* between the occurrence of peak temperatures on the soil surface and deeper soil layers is independent of

(a) Thermal conductance of soil
(b) Bulk density of soil
(c) Soil water content
(d) Soil color

[**Key:** i. b, ii. a, iii. c, iv. c, v. d, vi. b, vii. b, viii. c, ix. d, x. d]

References

Alli AA, Omofunmi OE (2021) A review of soil temperature under a controlled irrigation system. J Res in Forestry Wildlife Environ 13(1):50–59

Baver LD, Gardner WH, Gardner WR (1976) Soil physics. First Reprint. Wiley Eastern Ltd., New Delhi

Bottner P, Coûteaux MM, Vallejo VR (1995) Soil organic matter in Mediterranean-type ecosystems and global climatic changes: a case study—the soils of the Mediterranean basin. In: Moreno JM, Oechel WC (eds) Global change and Mediterranean-type ecosystems. Ecological studies. Springer, New York, NY, p 117

Chaudhary TN, Sandhu BS (1982) Soil temperature and plant growth. In Review of soil research in India. Part I. 12th Int. Congr. Soil Sci., New Delhi, India, pp 48–59

Christiansen MN (1963) Influence of chilling upon seedling emergence of cotton. Plant Physiol 38:520–522

Cooper AJ (1973) Root temperature and plant growth. Commonwealth Agr Bur, Slough

Frey S, Lee J, Melillo JM, Six J (2013) The temperature response of soil microbial efficiency and its feedback to climate. Nature Clim Change 3:395–398

Gao H, Shao M (2015) Effects of temperature changes on soil hydraulic properties. Soil Till Res 153:145–154

Glinski J, Lipiec J (1990) Soil conditions and plant roots. CRC Press, Boca Raton, FL

Grossnickle SC (2000) Ecophysiology of northern spruce species in the performance of planted seedlings. NRC–CNRC, NRC. Ottawa, ON, Research Press, pp 325–407

He H, Dyck MF, Horton R, Li M, Jin H, Si B (2018) Distributed temperature sensing for soil physical measurements and its similarity to heat pulse method. Adv Agron 18:173–230

Kohnke H (1968) Soil physics. TMH edition. Tata McGraw-Hill Publ. Co. Ltd, New Delhi

Kramer PJ (1940) Causes of decreased absorption of water by plants in a poorly aerated media. Am J Bot 27:216–220

Kramer PJ (1969) Plant and soil water relations—a modern synthesis. McGraw-Hill, New York, NY

Lahti M, Aphalo PJ, Finer L, Lehto T, Leinonen I, Mannerkoski H, Ryyppo A (2002) Soil temperature, gas exchange and nitrogen status of 5-year-old Norway spruce seedlings. Tree Physiol 22(18):1311–1316

Maximov NA (1929) The plant in relation to water. Allen and Unwin, London, pp 83–87

McMichael BL, Burke JJ (1998) Soil temperature and root growth. HortScience 33(6):947–951

Nielsen KF (1974) Roots and root temperature. In: Carson EW (ed) The plant and its environment. Univ Press of Virginia, Charlottesville, pp 293–335

Onwuka B, Mang B (2018) Effects of soil temperature on some soil properties and plant growth. Adv Plants Agric Res 8(1):34–37

Ruter JM, Ingram DL (1992) High root zone temperatures influence RuBisCo activity and pigment accumulation in leaves of *Rotundifolia holly*. J Am Soc Hortic Sci 117(1):154–157

Sachs J (1875) Geschichte Der Botanik Vom 16. Jahrhundert Bis 1860. München: Oldenbourg

Schipper LA, Hobbs JK, Rutledge S, Arcus VL (2014) Thermodynamic theory explains the temperature optima of soil microbial processes and high Q_{10} values at low temperatures. Glob Chang Biol 20:3578–3586

Sharma PK, De Datta SK (1991) Thermal characterization of a tropical rice soil in relation to puddling, flood-water depth and percolation rate. Soil Tech 4:167–175

Singh M, Sarkar B, Biswas B, Bolan NS, Churchman GJ (2017) Relationship between soil clay mineralogy and carbon protection capacity as influenced by temperature and moisture. Soil Biol and Biochem 109:95–106

Starr G, Neuman DS, Oberbauer F (2004) Ecophysiological analysis of two arctic sedges under reduced root temperatures. Physiol Plant 120:458–464

Wall DA, Stobbe EH (1984) The effect of tillage on soil temperature and corn (Zea mays L.) growth in Manitoba. Can J Plant Sci 64(1):59–67

Wanjura DF, Buxton DR, Stapleton HN (1970) A temperature model for predicting initial cotton emergence. Agron J 62:741–743

Chapter 8
Soil Strength and Plant Growth

Soil compaction is an indication of poor soil structure. It is a form of physical degradation of soil which has serious agricultural (crop quality and productivity), economical (energy requirement and cost of tillage), and environmental consequences (carbon recycling, soil erosion, environment pollution, etc.). It involves a complex interaction of soil, crops, machinery, and weather. Soil compaction, although needed in foundation engineering where soils are artificially compacted by rolling, tamping, vibrating, or compressing using suitable machines, is undesirable in agricultural lands. Soil compaction leads to soil densification, increase in soil strength, loss of total and air-filled porosity, reduction in water and air permeability, decrease in biological activity, and loss of soil productivity. It strongly influences soil-water-air-temperature relationships. It has an adverse effect on crop emergence, root growth and plant development, and nutrient and water uptake by crop plants. The net result is loss of crop yields. Several natural processes and anthropogenic activities are responsible for compaction of agricultural lands. Mechanized farming is the major factor responsible for compaction of agricultural lands in many regions of the world. Estimates indicate that around 68 million hectares of arable land has been degraded by soil compaction (Liu et al. 2022). Although severe soil compaction negatively affects soil properties, crop yields, and environment, moderate soil compaction sometimes is beneficial for yield increases by way of improving seed-soil and root-soil contacts. A knowledge and understanding of processes involved in soil compaction are essential in avoiding soil compaction as well as ameliorating compacted soils. In this chapter, we shall study the causes of soil compaction, its evaluation, management, and impact on plant growth.

P. K. Sharma, S. Kumar, *Soil Physical Environment and Plant Growth*,
https://doi.org/10.1007/978-3-031-28057-3_8

8.1 Important Terminology

8.1.1 Soil Densification

Soil densification refers to "the process of decrease in the volume of a unit mass of soil," i.e., increase in soil bulk density.

8.1.2 Soil Compression

Soil compression refers to "the process of soil densification due to the externally applied load or changes in the internal water pressure."

8.1.3 Soil Consolidation

Soil consolidation may be defined as "the gradual densification of a *saturated* soil under a continuously acting *static load*." It is soil compression under saturated conditions.

8.1.4 Soil Compaction

Soil compaction may be defined as "the densification of an *unsaturated* soil due to externally applied *dynamic load*." It is soil compression under unsaturated conditions.

8.1.5 Soil Compressibility

Soil compressibility is defined as "the resistance to change in volume (or bulk density) of soil when it is subjected to a mechanical load."

8.1.6 Soil Compactability

Soil compactability is "the difference between the initial bulk density and the maximum bulk density to which a soil can be compacted by a given amount of energy at defined water content."

8.2 Hardpans

A *hardpan* refers to "a dense or cemented soil horizon having low air and water permeability and obstructing root penetration." The hardpans may be of *genetic* origin (i.e., formed due to natural processes) or may be artificially *induced* due to anthropogenic activities. The examples of genetic hardpans include claypans, duripans, fragipans, etc. *Claypans* are formed due to the illuviation of clay from the A horizon into the B horizon. *Fragipan* is a brittle and compact subsurface horizon formed due to swelling/shrinkage and wetting/drying processes of soil, accumulation of amorphous silicates, illuviation of clay on grain contacts, overburden of ice mass, etc. *Duripans* are formed due to the precipitation of amorphous silica in subsurface horizons. The artificially induced hardpans, known as plow pans, plow soles, or traffic pans, result due to soil compaction caused by various farm operations. They are of concern in agricultural lands.

8.3 Soil Compression, Consolidation, and Compaction

Soil compaction and consolidation are two processes leading to soil densification, while soil compression is the umbrella term for both the processes.

8.3.1 Soil Compression

Soil compression is a process of densification of soil, i.e., decrease in the volume of soil per unit mass or increase in the bulk density of soil. Whenever a soil is subject to any kind of load, static or dynamic, there is a tendency of soil to undergo densification. During the process of soil compression, air and water are expelled from the pore space, and the soil particles reorient themselves into a closer packing. Depending on the soil water content, different processes may occur during compression. In extremely dry conditions, soil particles reorient themselves into closer packing, with the expulsion of air. In a saturated soil, on the other hand, compression may occur due to expulsion of pore water. In intermediate soil moisture conditions, between dryness and saturation, compression occurs due to expulsion of air and water. The compression of soils under saturated conditions is called *consolidation* and unsaturated conditions called *compaction*.

8.3.2 Soil Consolidation

Soil consolidation is a process of the gradual decrease in the volume of voids with the consequent increase in bulk density of a saturated soil under the action of a continuously acting static load over a period of time. It is a process in which saturated soil is compressed under the steady and static loads, such as landfill under its own load or load of structures or buildings, continuously acting over a period of time. The reduction in soil volume takes place by expulsion of water from voids. The soil particles get reoriented and packed together more tightly. The magnitude of consolidation depends on the compressibility of soils. The *compression index* of soil is determined with an *oedometer test.*

8.3.3 Soil Compaction

Soil compaction is a process of reduction in the volume per unit mass of an unsaturated soil due to externally applied dynamic loads. Soil compaction under a dynamic load primarily occurs due to the expulsion of air and water in the soil pores and the reorientation of soil solids into closer packing. Compression of soil solids, water, and air is practically nil or minimal, if at all. Soil compaction is important in agriculture, while soil consolidation is important in civil engineering.

8.4 Compaction of Agricultural Lands

The agricultural lands may experience soil compaction due to natural forces or anthropogenic activities. Compaction may occur on the soil surface (called *surface soil compaction*) or below the plow layer, i.e., below the normal depth of tillage operations (called *subsoil or subsurface compaction*).

The *surface soil compaction* generally refers to surface crusting, which is caused by the destruction of soil aggregates by raindrop impact, runoff, surface water stagnation, or tillage. Use of lighter farm machines and livestock trampling may also cause surface compaction. Zero tillage without proper organic residue management may also lead to surface compaction.

The *subsurface compaction* may be tillage induced or wheel traffic induced. The tillage-induced compaction is caused when soil is tilled repeatedly at the same depth using disc plow, moldboard plow, or sweep-type tools. The wheel traffic-induced compaction is caused by the axle weight load on the soil. Subsurface compaction occurs below the tillage depth. Subsurface compaction is more difficult to eliminate as compared to surface compaction.

8.4.1 Natural Processes of Soil Compaction

Following are a few examples of soil compaction due to the natural processes:

- Surface soil compaction (i.e., soil crusting) due to the beating and dispersive action of raindrops.
- Formation of claypans in subsoil due to the accumulation of clay; the clay accumulation may be depositional or due to illuviation of clay migrating from surface layers.
- Formation of fragipans or duripans at the junction of two distinctly different layers in subsoil due to cementation by illuvial silica.

8.4.2 Anthropogenic Activities Causing Soil Compaction

Several farm activities are responsible for the compaction of agricultural lands. Important ones include frequent and excessive traffic of farm equipment and animals, uncontrolled and unscientific grazing by animals, less crop diversification, reduction in soil organic matter content, intensive tillage, and tillage at high moisture contents. Important anthropogenic activities leading to soil compaction are briefly discussed below:

8.4.2.1 Farm Mechanization

High mechanical loads and contact pressures that create noticeable ruts in the topsoil are the primary causes of compaction of agricultural lands. Fragile, loose, and wet soils are more susceptible to traffic-induced compaction. Mechanized farming performs most of the field operations from sowing to harvesting mechanically by using heavy machines. The machines which induce stress more than the soil strength cause soil compaction. Soil below the plow layer gets compacted. The subsoil compact layers (2.5–5.0 cm thick) are called *plow pans*, *plow soles*, *tillage pans*, *traffic pans*, etc. The degree and extent of soil compaction depend on:

(a) The machine characteristics (axle load, pressure and vibrations of the machinery, type of machine used, tire dimensions, inflation pressure of tires, etc.)
(b) The mode of operation of farm machines (frequency, number of passes and speed of machines, vibration applied, soil-tire interaction, wheel slippage, etc.)
(c) Soil properties (soil texture, structure, bulk density, soil water content, soil strength, organic matter content, depth and nature of surface litter, etc.)

Vehicles and machines with higher axle load operated at high moisture contents compact soils to deeper depths, and the subsoil compaction is almost permanent. Coarse-textured soils are less susceptible to compaction than fine-textured soils

(Horn and Lebert, 1994). Soil aggregation and organic matter decrease the susceptibility of soils to compaction (Baumgartl and Horn, 1991; Nawaz et al. 2013).

8.4.2.2 Puddling

Subsoil compaction is commonly observed in typical rice soils, especially those which are under rice-rice cropping, due to intensive wet tillage called puddling for planting of rice. Puddling destroys soil structure. Upon drying, the puddled soils shrink and get compacted causing surface compaction. Sometimes, soils become so dense and hard that it becomes difficult to till them with common tillage implements. Heavy-duty tractors are needed to plow such soils. Puddled soils also exhibit subsoil compaction due to intensive use of tractors, puddlers, and other farm machinery. Soil strength (as measured with soil penetrometer) of >3 MPa has been observed in rice fields in several south and Southeast Asian countries (JC O'Toole, personal communication, cited by Dexter and Woodhead, 1985).

8.4.2.3 Trampling by Livestock

Livestock farming on meadows and agricultural lands results in significant livestock trampling. The compaction caused due to trampling by livestock is usually limited to the surface layers (top 2–5 cm depth). The intensity of grazing is more important than the period of grazing.

8.5 Trafficability and Workability of Agricultural Soils

The trafficability and workability of soils refer to their ability to support farm mechanization. The *trafficability* may be defined as "the ability of soil to support agricultural traffic without undergoing physical degradation." It determines soil's fitness to allow safe mobility of farm machinery in the field. The *workability* may be defined as "the ability of soil to support tillage operations." It is the ability of soil to be tilled. It determines the feasibility of performing tillage operations in the field without damaging soil structure.

The trafficability and workability determine the operational feasibility of farm traffic and machinery, and the resulting physical status of soil. Both are strongly dependent on soil texture and soil moisture content. Weather conditions, land-use pattern, and management practices also significantly influence both the parameters. If the rules of trafficability and workability are not followed judiciously, it may lead to physical degradation of soils including soil compaction, threatening the sustainability of the crop production.

8.5.1 *Assessment of Trafficability and Workability of Soils*

The trafficability and workability of soil depend primarily on its moisture content. The optimum soil moisture content which supports these processes can be assessed by using Atterberg consistency limits, proctor moisture content, or soil moisture characteristics.

Soil strength in general is an important soil parameter determining trafficability and workability of soils. Soil strength affects the bearing capacity of soils, traction parameters of vehicles, and energy required for tillage. Soil strength depends on soil moisture content and bulk density of soil. Soil texture is also an important factor. Excessive water contents make soils soft and slippery, and increase their stickiness, thereby restricting the mobility of farm traffic. Firm soil matrix supports load and shear stress of farm machines to avoid soil compaction. Strong interaction also exists between the type of farm machine to be used and the moisture status of soil.

8.5.1.1 Atterberg's Consistency Limits

The moisture range for the best workability may be determined from Atterberg's consistency limits. As a thumb rule, too wet or too dry soil conditions are not good for tillage and thus indicate poor workability. Tillage in dry soils, i.e., at hard consistency, ends up in coarse seedbed with cloddy tilth. Tillage of wet soils, i.e., at plastic limit, results in plastic deformation of soil, while at liquid limit results in a soft soil puddle which becomes compact and hard upon drying. Best workability is achieved at moisture contents between plastic limit (PL) and shrinkage limit (SL) of soil. Best workability for some soils may be at gravimetric moisture content around 0.9 times the plastic limit (Utomo and Dexter, 1981). Tillage at this moisture content produces larger proportion of micro-aggregates (Ojeniyi and Dexter, 1979). The difference between PL and SL is known as *friability range* or *friability index* (FI). Soils with high value of FI remain friable over a wider moisture range. A strong interaction occurs between soil texture and soil moisture for the friability range.

Thomasson (1982) used the Atterberg's lower plastic limit (LPL) as the threshold limit of workability. Kretschmer (1996) (cited by Müller et al. 2011) used *consistency index* (I_c) to determine soil workability. The I_c is computed as follows:

$$I_c = (\mathrm{UPL} - \mathrm{w}) / (\mathrm{UPL} - \mathrm{LPL}) \tag{8.1}$$

where UPL is the upper plastic limit, LPL the lower plastic limit (i.e., *shrinkage limit*), and w the actual field water content, with the UPL, LPL, and w being expressed as kg water/kg dry soil. A consistency index of $I_c = 0.75$ is considered as an arbitrary limit of workability.

The concept of Atterberg's consistency limits is a useful index of workability under field conditions, but in case of cohesive soils only. The concept is not applicable in non-cohesive soils which lack plasticity.

8.5.1.2 Proctor Moisture Content

Proctor moisture content, which determines the relationship between soil moisture content and maximum compaction at a defined energy impact, can be used to assess the trafficability and workability of soils. Each soil has a definite moisture content at which maximum bulk density is achieved with a given energy impact. It is called the proctor moisture content. Soils are best tilled at moisture contents below the proctor moisture content.

8.5.1.3 Soil Moisture Characteristics

Approximate estimation of trafficability and workability of soils can also be assessed from the soil moisture characteristics. The field capacity or matric potential between −5.0 and −30 kPa gives an approximate acceptable condition of performing field operations (Müller et al. 2011).

8.6 Evaluation of Soil Compaction

Soil compaction may be observed directly in the field or by studying soil and plant samples in the laboratory or by remote sensing. Soil compaction under field conditions may be detected by careful visual observations (qualitative assessment) or systematic measurements of changes in soil and crop characteristics (quantitatively), viz. soil structure, soil moisture, soil color, penetrometer resistance, air and water permeability, surface water ponding, waterlogging, and root and crop characteristics. Quantitative assessment gives a better understanding of correlation between plant growth and soil compaction.

8.6.1 Visual Characteristics of Soil Compaction

Visual observations of soil conditions and crop performance may give a good idea of soil compaction. Surface compaction is easier to identify with such observations than the subsurface compaction. Following observations may help in identifying soil compaction:

- Formation of surface crusts
- Localized ponding of water in the field during rains
- Formation of waterlogging zones
- Formation of zones of early drying of crop plants due to shallow rooting depth
- Early drying up of plants at the end of the cropping season
- Signs of moisture stress on plants relatively quickly after the rainfall or irrigation
- Delayed flowering particularly in broadleaf crops such as lupin
- Deformed and bent plant roots, stunted plant growth, and low crop yields
- Poor root growth with swollen tips, root growth confined to spaces between large soil blocks, or horizontal root growth over the compact soil layers
- Poor plant growth and yield in wheel tracks
- Difficulty in pushing metallic hand probes (8–10 mm dia steel rods) into soil
- Increased draft force of the tractor during tillage
- Formation of dense soil clods upon deep tillage
- Formation of fractured blocky or platy structure in horizontal orientation and very few visible soil pores in subsoil

8.6.2 Quantitative Assessment of Soil Compaction

Soil compaction under field conditions can be characterized by making measurements in the field or laboratory.

8.6.2.1 Soil Strength

Soil strength, also known as *soil mechanical impedance* or *soil penetration resistance*, is an important and most used index of soil compaction. Soil strength may be defined as "the capacity of soil to resist or endure an applied force against any deformation." The resistance is both to volumetric compression and linear deformation.

Soil strength under field conditions is commonly measured with cone penetrometers because of their fast and simple handling and reliability. The penetrometer probe imitates the plant root. A cone *penetrometer* is manually pushed into the soil at a steady speed of around 30 mm/s, and the force required to push the penetrometer into soil is called *soil penetration resistance* or *cone index*. It is recorded as kilopascals (kPa) or megapascals (MPa). Soil penetration resistance is directly related to the bulk density and indirectly to the soil moisture content. At a given bulk density, soil strength decreases with increasing moisture content, while at a given moisture content, soil strength increases with increasing bulk density.

Soil strength with cone penetrometers should be measured when soil is at or around field capacity. Cone penetration resistance should only be compared between soils having the same texture, structure, and moisture content. Also, the cone penetration data in stony and gravely soils should be interpreted with care. Soil strength is a better index than bulk density in interpreting the effects of soil compaction on plant growth.

Table 8.1 Optimum and root growth-restricting bulk density in relation to soil texture (Nyeki et al. 2017)

Soil texture	Bulk density optimum for root growth (g/cc)	Bulk density-restricting root growth (g/cc)
Sand, loamy sand	<1.60	>1.80
Sandy loam, loam	<1.40	>1.80
Sandy clay loam, clay loam	<1.40	>1.75
Silt, silt loam	<1.30	>1.75
Silty clay loam	<1.40	>1.65
Sandy clay, silty clay	<1.10	>1.58
Clay	<1.10	>1.47

Soil penetration resistance can also be estimated using data on bulk density and volumetric moisture content. One such model developed by Jakobsen and Dexter (1987) was validated by Liu et al. (2022) to a greater accuracy under field conditions. According to this model, soil penetration resistance (Q) is computed as a function of bulk density (ρ_b, Mg/m^3) and volumetric moisture content (θ_v, m^3/m^3) as follows:

$$Q = e^{a\theta v + b\rho b + c} \tag{8.2}$$

where a, b, and c are constants.

8.6.2.2 Bulk Density

High bulk density is a direct measure of soil compaction. Bulk density is negatively correlated with root growth (Table 8.1). Bulk density values critical for root growth and development vary with the soil texture as well as crop species. In general, bulk densities that impede root growth are 1.55 Mg m^{-3} for clay loams, 1.65 Mg m^{-3} for silt loams, 1.80 Mg m^{-3} for sandy loams, and 1.85 Mg m^{-3} for loamy fine sands.

8.6.2.3 Pore Size Distribution

Soil compaction modifies pore size distribution. Microporosity increases at the cost of macroporosity due to soil compaction. The significance of pore size distribution on air-water transmission properties of soil and on root growth is described in Chap. 5 (Sect. 5.11). The macropores are essential for maintaining water and air conductivity in soils. They provide drainage and are essential for good air exchange during wet periods. Macropores are also needed for root growth. Douglas and Crawford (1993) reported reduction in the volume of macropores to less than half of that in the non-compacted soil with consequent reduction in air permeability and infiltration rate (Table 8.2). Micropores influence hydraulic conductivity and ionic diffusion in

Table 8.2 Effect of soil compaction on macropores, air permeability, and infiltration rate (Douglas and Crawford, 1993)

Compaction treatment	Macropore volume (ft^3/ft^3)	Air permeability (mm^2)	Infiltration rate (inch/h)
Compacted	0.044	1.0	0.25
Non-compacted	0.119	55.0	1.06

soils. They are critical for water retention and help a crop go through dry periods with minimal yield loss. Different laboratory methods are available for determining pore size distribution, viz. pressure plate or pressure membrane apparatus and mercury intrusion porosimetry. Each method has its own advantages and limitations. Pressure plate or pressure membrane apparatus is commonly used to determine pore size distribution in undisturbed soil core samples.

8.7 Effect of Compaction on Soil Quality

Soil particles come to a closer packing during soil compaction. There is reduction in soil porosity leading to a reduction in space available for air and water and change in thermal properties. Gürsoy (2021) has reviewed and summarized the effects of soil compaction on soil properties, environment, and plant growth. The soil quality is affected adversely in a number of ways:

- Destruction of soil structure; crushing of soil aggregates due to compaction forces leading to a more massive structure with fewer natural voids
- Reduction in total porosity, pore continuity, and connectivity
- Reduction in air and water transmission capacity of soil leading to surface water ponding, waterlogging, poor drainage, and poor soil aeration
- Reduction in oxygen concentration in soil air, which may lead to soil hypoxia, increased denitrification, and reduced rate of decomposition of organic matter and N mineralization
- Increase in the surface runoff and soil erosion potential
- Reduced biological activity—population and activity of nematodes and earthworms are reduced, while bacteria, fungi, and protozoa are less affected (effect may be negligible)
- Reduced soil productivity

8.8 Effect of Soil Compaction on Plant Growth

Soil compaction has become a major environmental threat to plant growth and development. The threat is becoming more and more serious with time under mechanized farming due to progressive increase in the size and weight of farm equipment (Soane and Van Ouwerkerk, 1998; Keller et al. 2019) (Fig. 8.1).

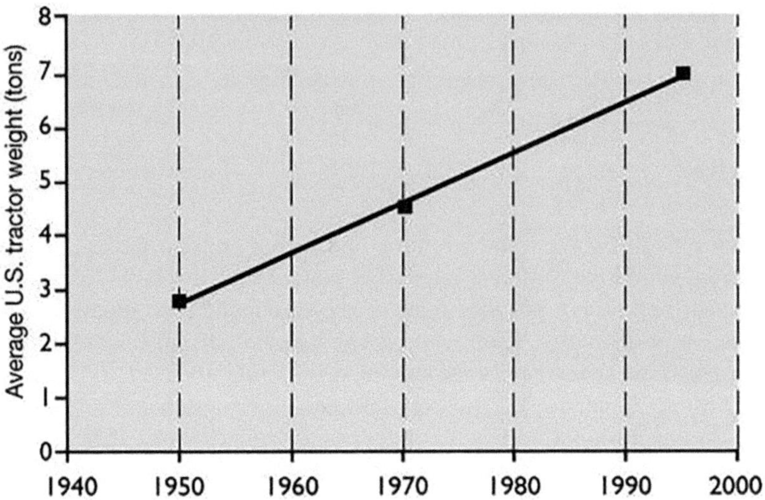

Fig. 8.1 Increase in tractor weight with time since the 1950s (Soane and Van Ouwerkerk, 1998)

Soil compaction adversely affects crop production in a number of ways:

- Reduction in seedling emergence by surface compaction (soil crusting)
- Restricted root growth and development
- Restricted soil volume explored by plant roots for water and nutrients
- Reduced availability of plant nutrients
- Reduction in nutrient and water absorption by plant roots
- Reduction in the enzymatic activity and associated metabolic processes in plant system
- Reduction in nodulation in leguminous plants
- Increased root-rot diseases
- Reduction in crop yield

8.8.1 Soil Compaction and Root Growth

Root growth and development (cell development and elongation, root length, diameter and surface area, root architecture, etc.) are significantly influenced by soil compaction. High soil penetration resistance and poor soil aeration reduce root elongation rate, delay the initiation of lateral roots, and produce shallow root systems leading to restricted soil exploration and limited accessibility of plants to water and nutrients, and ultimately low crop yields. The dominant physical stress for growing roots in compact soils is high soil penetration resistance under dry conditions and poor soil aeration under wet conditions. It suggests the role of weather conditions (dry or wet year) on the interaction between soil compaction and root growth.

Roots grow in soil pores which are mostly smaller in diameter than the roots. So to grow, the roots must deform the soil. Growing roots exert radial (or volumetric) and longitudinal (or axial) pressure, collectively called *root pressure*, to counterbalance the soil penetration resistance (Sect. 5.11.2). As long as the root pressure is more than the soil penetration resistance, roots continue to grow. If penetration resistance is more than the root pressure, root growth is restricted.

A strong interaction exists among root elongation, soil penetration resistance, and soil moisture status. Soil penetration resistance increases with decreasing moisture content. In a survey of 19 soils, ranging in texture from loamy sand to silty clay loam, Bengough et al. (2011) observed that while at a matric potential of −10 kPa around 10% soils exhibited penetration resistances of >2 MPa, at −200 kPa matric potential, the number increased to nearly 50%. It suggests that while interpreting the effects of soil compaction on root growth, the soil moisture status should also be critically considered.

An inverse relationship exists between soil penetration resistance and root growth (Fig. 8.2). In general, restriction on root growth of most of the cereal crops starts at penetration resistance of around 1.5 MPa when the soil moisture content is at field capacity. Root growth is severely restricted at penetration resistance of ≥2.5 MPa. Willatt (1986) observed decline in rooting depth as well as root length density of barley in 0–30 cm soil layer due to soil compaction with six passes of tractor on the soil surface.

Roots grow thicker in compacted soils. Kirby and Bengough (2002) reported 40% and 60% thicker pea roots at 1.0 and 2.0 MPa penetration resistances, respectively, as compared to roots growing at 0.7 MPa penetration resistance in a sandy loam soil. Root thickening was comparatively less in a clay loam soil; it was 10% and 20%, respectively, at 1.0 and 2.0 MPa penetration resistance than at 0.7 MPa penetration resistance. A growing root experiences maximum stress at the tip of root cap (Kirby and Bengough, 2002), and root thickening is a mechanism of decreasing this stress and reducing the risk of root buckling (Kirby and Bengough, 2002;

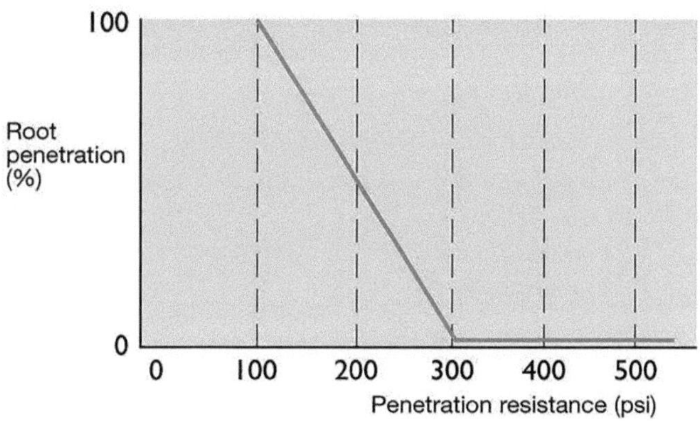

Fig. 8.2 Root penetration in relation to soil penetration resistance (1 psi = 6.89476 kPa) (Taylor et al. 1966)

Chimungu et al., 2015). Root penetrability depends more on the root anatomical phenes than on the root diameter per se (Chimungu et al. 2015).

8.8.2 Soil Compaction and Nutrient Absorption

Nutrient absorption and utilization by crop plants depend on the availability of nutrients in the rhizosphere, which in turn depends on the soil exploration by root system, ionic mobility in soils, and enzymatic activity within the plant system. For roots to access nutrients in soil, either the nutrients should move to the root surface through mass flux (most of the N, S, Ca, and Mg, and to some extent K) and/or diffusion (most of the P and K, and micronutrients) or the roots themselves intercept the nutrients in soil. The relative significance of root interception, mass flow, and diffusion in ion transport to plant roots is shown in Table 8.3. Nutrient access for N, P, K, and S through root interception is generally $\leq 4\%$ of the available nutrients in soil. Mass flux is important in case of N, Ca, Mg, and S, while diffusion is important in case of P and K for making them available at root surface.

Soil compaction affects root interception by restricting root growth, while mass flow and diffusion are affected by change in pore size distribution. Soil compaction decreases macroporosity and increases microporosity. Decrease in macroporosity decreases the mass flux. Increase in microporosity initially increases anionic diffusion due to their increased negative adsorption. But above certain critical level, soil compaction decreases ionic diffusion due to reduction in the effective area available for diffusion.

Several studies report reduction in nutrient absorption at high compaction levels. In one greenhouse study, increase in soil penetration resistance from 1.5 MPa to 5.2 MPa reduced absorption of N, K, Mg, and Na by 13.5, 51.4, 50.4, and 51.5% by maize seedlings; the absorption of Ca and P was maximum at the intermediate level of soil penetration resistance of 3–4 MPa (Olubanjo and Yessoufou, 2019). Lipiec and Stepniewski (1995) reported significant reduction in P and K uptake in compact soils. Parlak and Parlak (2011) reported decrease in the absorption of N, Ca, and Mg; increase in Fe, Mn, and Zn; and irregular absorption pattern of P and K by

Table 8.3 Relative contribution of root interception, mass flow, and diffusion in nutrient uptake by maize roots (Havlin et al. 2005)

	Root interception	Mass flow	Diffusion
Nutrient	Percent supply		
N	1	99	0
P	2	4	94
K	2	20	78
Ca	12	88	0
Mg	27	73	0
S	4	94	2

vetch, common vetch, Italian ryegrass, and barley when bulk density of a clay loam soil increased from 1.3 through 1.4, 1.5, to 1.6 g/cm^3. Low oxygen levels in compacted soils are also responsible for significant reduction in the absorption of Mg.

The N availability in a compacted soil is affected in a number of ways:

- Increased denitrification and reduced mineralization of organic matter due to poor drainage in compacted soils
- Decreased leaching losses of nitrates
- Increased loss of surface-applied nitrogen fertilizers
- Reduced diffusion/mass flow of nitrates and ammonium ions to the plant roots

Reduction in N mineralization rate by 30% and increase in denitrification rate by 20% due to soil compaction were reported by Douglas and Crawford (1993). These workers also observed that N-use efficiency in compacted soil was low and to achieve the same dry matter production as in non-compacted soil the N dose had to be doubled.

Soil compaction also impacts the ionic accumulation in plant tissues by influencing the activity of certain enzymes. Wang et al. (2019) observed increased activities of pyruvate kinase and phosphofructokinase enzymes in soybean seedling roots at moderate-level soil compaction, resulting in accumulation of P, K, Mg, Ca, and other elements in soybean seedlings. It resulted in increased number of fibrous upper roots, but reduced root length and inhibited plant growth. High-level soil compaction inhibited the accumulation of P, K, Mg, Mn, Fe, Cu, and Zn but increased the accumulation of Ca due to decreased activities of isocitrate dehydrogenase and cytochrome oxidase enzymes. It resulted in decreased root cell size, blurred root cell boundaries, and inhibition of plant growth. Overall, soil compaction inhibited plant growth.

8.8.3 Soil Compaction and Crop Yield

Soil compaction modifies soil physical, chemical, and biological properties in such a way that they become unfavorable for plant growth and yield. Compact soils are poor crop yielders. High soil penetration resistance and low air and water permeability in compact soils lead to restricted root growth, reduced nutrient absorption, accumulation and utilization by plants, increased infections of root-rot diseases, etc., which result in low crop yields. Obour and Ugarte (2021) while analyzing the results of 51 published articles reported significant declines in the yields of corn, wheat, barley, and soybean due to soil compaction caused by mechanical farming. Similarly, Keller et al. (2019) associated the observed yield declines in the major food crops in many European countries with the soil compaction under mechanized farming. Olubanjo and Yessoufou (2019) reported decline in maize grain yield by 18% due to increase in soil penetration resistance from 1.5 to 5.2 MPa.

Crop yields are generally low at very low or very high bulk densities. Moderate soil compaction is desirable for better seed germination, seedling emergence, crop

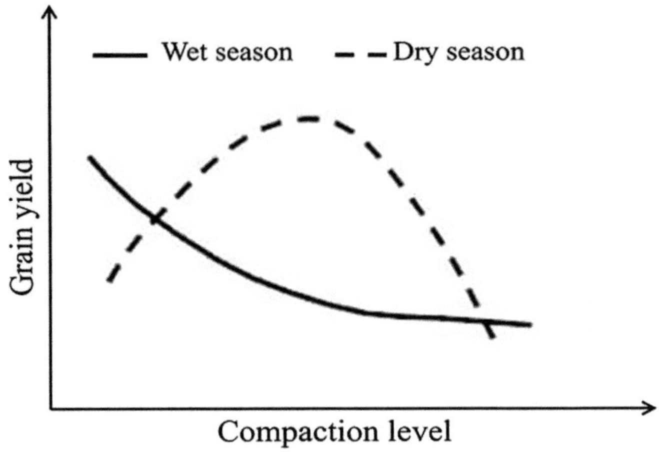

Fig. 8.3 Effects of weather on crop yield response to compaction level

stand and yield due to optimum seed-soil contact, air-water relations, and nutrient availability (Sivarajan et al. 2018; de Moraes et al. 2020). But if compaction increases beyond certain critical level, crop yields decline. Weather conditions strongly influence the response of crop yields to compaction levels (Nyeki et al. 2017; Liu et al. 2022) (Fig. 8.3). In a dry year, moderate compaction improves seed germination, crop stand, and yield. Crop yields decline at high compaction levels due to restricted root growth, reduced nutrient absorption, and stunted plant growth. To sustain crop yields, proper water and nutrient management is needed. In wet years, crops in compact soils experience stresses like poor aeration, decreased nutrient availability, and increased risk of root diseases, leading to progressive decline in crop yields. Studies indicate that plants can tolerate soil compaction stress in a better way during dry year than during wet year (Voorhees, 1986).

Tillage can alleviate ill effects of compaction on crop yields. Surface compaction can be easily ameliorated while subsoil compaction is long lasting. Tillage may alleviate surface compaction effects within a year in sandy soils, while it takes more tillage passes and repeated freeze-dry cycles in fine-textured soils. Freezing-thawing and wetting-drying cycles have little to no effect on subsoil compaction in any type of soil. A multilocation field trial in the United States studied the effect of tillage after compaction on crop yield losses (Hakansson and Reeder, 1994). Average yield losses in the first year after tillage were around 15%, which were considered to be due to the residual effect of surface compaction. The yield losses reduced with time and stabilized around 3% within 10 years' period (Fig. 8.4). This yield loss was considered to be permanent and was assigned to the subsoil compaction.

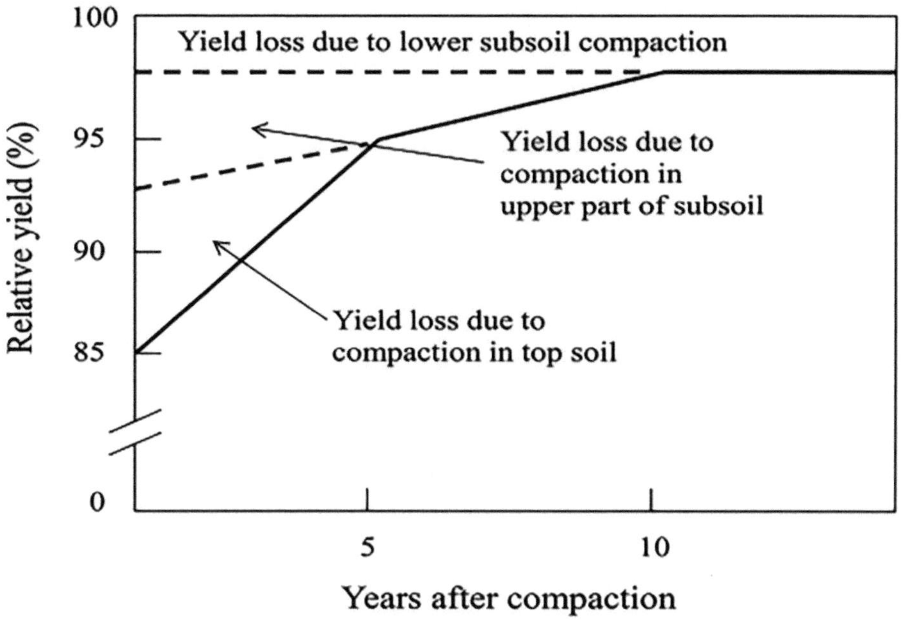

Fig. 8.4 Relative crop yield on compacted soil compared to non-compacted soil with moldboard plowing (Hakansson and Reeder, 1994)

8.9 Management of Soil Compaction

Soil compaction in agricultural fields can be managed in three ways:

(i) Avoid soil compaction
(ii) Ameliorate compacted soils
(iii) Alleviate stresses associated with soil compaction

The first step in the management of soil compaction is to properly diagnose the cause and problem of soil compaction, followed by adoption of short- and long-term management practices to prevent, ameliorate, or alleviate soil compaction-related stresses. It may also be noted that amelioration of subsoil compaction is much more expensive and less effective than loosening the compacted surface soils. Avoidance of subsurface compaction is therefore very important. Confinement of damage to specific strips in the field is a more realistic alternative than compaction avoidance. The following practices can avoid, delay, or prevent soil compaction (Alakkuku et al. 2003; Hamza and Anderson, 2005; Bengough et al. 2011; Idowu and Angadi, 2013; Nawaz et al. 2013; Kumar et al. 2018):

- Reduce pressure on soil either by decreasing axle load or by increasing the contact area of wheels with the soil. To avoid subsoil compaction, keep the single axle load of 4–6 tonnes and the tandem axle load of 8–10 tonnes for moist mineral soils (Alakkuku et al. 2003). The machine load should not exceed soil strength (i.e., bearing capacity of soil).

- Avoid trafficking the wet soils, i.e., avoid tillage at plastic limit or above; use farm machines; conduct field operations; and allow grazing of animals at optimal soil moisture content, i.e., moisture content in the friability range.
- Use tires of proper size and with proper inflation; the use of dual wheels, rubber tracks, and flotation tires also helps in avoiding soil compaction.
- Reduce the intensity and frequency of tillage. Use of larger equipment and adoption of minimum/zero tillage reduce the number of passes in the field.
- Reduce the intensity of animal grazing.
- Decrease trafficked area in the field by using machines having wide spacing between the wheels. Use controlled traffic by confining traffic to permanent traffic lanes. Avoid trafficking the area between the lanes.
- Increase soil organic matter through retention of crop and pasture residues on soil surface. Adopt conservation tillage. Avoid inversion type of tillage.
- Use crop rotations that include plants with deep and strong taproots capable of penetrating and puncturing the subsurface hardpans.
- Improve water-stable aggregation and soil structural stability through SOC buildup.
- Adopt conservation tillage to improve soil organic matter content, aggregation and more stable soil structure, and soil hydraulic properties.
- Maintain soil fertility to enable soils and crops to resist harmful compaction-related stresses.
- Use subsoiling, chiseling, or deep ripping for eliminating soil compaction, destroying hardpans, and ameliorating hard-setting soils due to traffic and puddling.

8.10 Question Bank

8.10.1 Short Questions:

(i) Define soil compression, compaction, and consolidation. What is the basic difference among the three processes? Which of the three processes is important in agriculture?

(ii) What is soil compaction? Briefly describe natural processes and anthropogenic activities responsible for soil compaction.

(iii) Define trafficability and workability of agricultural soils. Briefly describe the methods of evaluation of these two processes.

(iv) What is soil strength? How is it influenced by bulk density and moisture content of soil? Briefly describe the assessment of soil strength with soil penetrometer.

(v) Can soil compaction be evaluated through visual observations in the field? If yes, how? If no, why not?
(vi) How will you quantitatively assess soil strength?
(vii) What is the impact of soil compaction on
  (a) Soil properties
  (b) Plant growth
  (c) Root growth
  (d) Nutrient absorption by plants
  (e) Crop yield
(viii) How will you manage compacted soils for crop production?

8.10.2 *Briefly Explain Why?*

(i) Soil compaction is more important in agriculture than soil consolidation.
(ii) Soil compaction has become a serious issue due to increased mechanized farming.
(iii) Anthropogenic activities are primarily responsible for compaction of agricultural lands.
(iv) Puddling causes soil compaction.
(v) Noncompliance of trafficability and workability rules during crop production may lead to physical degradation of soils.
(vi) Soils should be tilled at moisture contents between plastic limit (PL) and shrinkage limit (SL) of soil.
(vii) Atterberg's consistency limit is a useful index of workability in cohesive soils only.
(viii) Soils are best tilled at moisture contents below the proctor moisture content.
(ix) Root growth and soil penetration resistance are negatively correlated.
(x) Soil compaction decreases nutrient absorption by plant roots.
(xi) Soil compaction has a significant effect on N availability in soils.
(xii) Compact soils are poor crop yielders.

8.10.3 *Fill in the Blanks:*

(i) The process of decrease in the volume of a unit mass of soil is known as ....
(ii) The densification of an unsaturated soil due to externally applied dynamic load is known as ....
(iii) The gradual densification of a saturated soil under a continuously acting static load is known as ....

(iv) The difference between the initial bulk density and the maximum bulk density to which a soil can be compacted by a given amount of energy at defined water content is known as ....
(v) The .... of soil is determined with an oedometer test.
(vi) The .... is formed due to the cementation by illuvial silica at the junction of two distinctly different soil layers.
(vii) The farm mechanization is the major anthropogenic activity causing soil ....
(viii) Coarse-textured soils are ...... susceptible to compaction than fine-textured soils.
(ix) The ability of soil to support agricultural traffic without undergoing physical degradation is known as the .... of soil.
(x) The difference between the moisture contents at plastic limit and shrinkage limit of soil is known as ....
(xi) Soils are best tilled at moisture contents .... the proctor moisture content.
(xii) Soil penetrometer is used to measure ....
(xiii) Soil penetration resistance .... with increasing bulk density and decreasing moisture content.
(xiv) The magnitude of bulk density critical for root growth .... with the fineness of soil texture.
(xv) Soil compaction .... microporosity and .... macroporosity.
(xvi) Mercury intrusion porosimetry technique is used to determine ....
(xvii) Nutrient access for NPK by roots through root interception is ....

[**Key:** i. soil densification, ii. soil compaction, iii. soil consolidation, iv. soil compactability, v. compression index, vi. fragipan or duripan, vii. compaction, viii. less, ix. trafficability, x. friability moisture range, xi. below, xii. soil strength/soil penetration resistance, xiii. increases, xiv. decreases, xv. increases, decreases, xvi. pore size distribution, xvii. $\leq$4%]

8.10.4 State Whether the Following Statements Are True (T) or False (F):

(i) Soil compaction, soil compression, and soil consolidation are synonymous terms.
(ii) Densification of saturated soils under static load is called soil compaction.
(iii) Densification of unsaturated soils under dynamic load is called soil compaction.
(iv) Densification of saturated soils under static load is called soil consolidation.
(v) During compaction, the reduction in soil volume takes place by expulsion of water from voids.

(vi) Soil crusting is an example of surface soil compaction.
(vii) Coarse-textured soils are less susceptible to compaction than fine-textured soils.
(viii) The trafficability and workability of soils refer to their ability to support farm mechanization.
(ix) The moisture range for the best trafficability may be determined from Atterberg's consistency limits.
(x) Soils are best tilled over a part of friability range.
(xi) Soils are best tilled at the proctor moisture content.
(xii) Delayed flowering particularly in broadleaf crops is an indication of soil compaction.
(xiii) Formation of dense soil clods upon deep tillage is an indication of soil compaction.
(xiv) There is indirect relationship between soil strength and bulk density.
(xv) There is indirect relationship between soil strength and moisture content.
(xvi) Bulk density is a direct measure of soil compaction.
(xvii) Bulk density-restricting root growth increases with the fineness of soil texture.
(xviii) Micropores influence hydraulic properties of soil, while macropores are essential for root growth.
(xix) Soil compaction-induced oxygen stress may lead to denitrification in soils.
(xx) The root growth pressure must counterbalance the soil penetration resistance for the roots to grow.
(xxi) Soil compaction has little or no effect on nutrient absorption by roots.
(xxii) Plat roots access P and K in soils primarily through diffusion.
(xxiii) Soil compaction retards N mineralization but increases denitrification.

[**Key:** i. F, ii. F, iii. T, iv. T, v. F, vi. T, vii. T, viii. T, ix. F, x. T, xi. F, xii. T, xiii. T, xiv. F, xv. T, xvi. T, xvii. F, xviii. T, xix. T, xx. T, xxi. F, xxii. T, xxiii. T]

8.10.5 Multiple-Choice Questions:

(i) Soil densification due to externally applied dynamic load or static load irrespective of soil water content is known as

(a) Soil compaction
(b) Soil compression
(c) Soil consolidation
(d) Soil deformation

(ii) Which one of the following hardpans is not formed due to natural processes:

(a) Claypan
(b) Fragipan
(c) Duripan
(d) Plow pan

(iii) Which one of the following hardpans is formed due to natural processes:

(a) Claypan
(b) Plow pan
(c) Traffic pan
(d) Plow sole

(iv) During the process of soil compaction, there is

(a) Expulsion of water from the pore space
(b) Expulsion of air from the pore space
(c) Expulsion of water and air from the pore space
(d) None of the above

(v) Subsoil compaction may be

(a) Tillage induced
(b) Traffic induced
(c) Both tillage and traffic induced
(d) Livestock trampling induced

(vi) The ability of soil to support tillage operations is known as

(a) Dependability
(b) Trafficability
(c) Sustainability
(d) Workability

(vii) The following value of consistency index (I_c) is considered as an arbitrary limit of workability (Müller et al. 2011):

(a) 0.25
(b) 0.50
(c) 0.75
(d) 1.00

(viii) Soils are best tilled at moisture contents

(a) Below the proctor moisture content
(b) At the proctor moisture content
(c) Above the proctor moisture content
(d) None of the above

(ix) The best index of soil compaction with respect to plant growth is

(a) Bulk density
(b) Soil penetration resistance
(c) Total porosity
(d) Pore size distribution

(x) Soil penetration resistance is measured in the field by

(a) Porometer
(b) Oedometer
(c) Penetrometer
(d) Modulus of rupture test

(xi) Restriction on root growth of most of the crops starts at the following soil penetration resistance measured at field capacity:

(a) 0.5 MPa
(b) 1.5 MPa
(c) 2.5 MPa
(d) 3.5 MPa

(xii) An inverse relationship exists between

(a) Soil strength and bulk density
(b) Soil strength and denitrification
(c) Soil strength and root rot diseases
(d) Soil strength and nutrient absorption by plant roots

(xiii) The major contributory process in making P and K ions available in the rhizosphere is

(a) Root interception
(b) Mass flow
(c) Diffusion
(d) Vapor flux

[**Key:** i. b, ii. d, iii. a, iv. c, v. c, vi. d, vii. c, viii. a, ix. b, x. c, xi. b, xii. d, xiii. c]

References

Alakkuku L, Weisskopf P, Chamen WCT, Tijink FGJ, Van der Lindane JP, Pires S, Sommer C, Spoor G (2003) Prevention strategies for field traffic-induced subsoil compaction: a review. Part 1. Machine/soil interactions. Soil Tillage Res 73:145–160

Baumgartl T, Horn R (1991) Effect of aggregate stability on soil compaction. Soil Tillage Res 19:203–213

Bengough AG, McKenzie BM, Hallett PD, Valentine TA (2011) Root elongation, water stress and mechanical impedance: a review of limiting stresses and beneficial root tip traits. J Exp Bot 62(1):59–68

Chimungu JG, Loades KW, Lynch JP (2015) Root anatomical phenes predict root penetration ability and biomechanical properties in mays (*Zea mays*). J Exptl Bot 66(11):3151–3162

De Moraes MT, Debiasi H, Franchini JC, Mastroberti AA, Levien R, Leitner D, Schnepf A (2020) Soil compaction impacts soybean root growth in an Oxisol from subtropical Brazil. Soil Tillage Res 200:104611

Dexter AR, Woodhead T (1985) Soil mechanics in relation to tillage, implements and root penetration in lowland soils. In: Soil physics and rice. Int Rice Res Inst, Los Banos, Philippines

Douglas JT, Crawford CE (1993) The responses of a ryegrass sward to wheel traffic and applied nitrogen. Grass Forage Sci 48:91–100

Gürsoy S (2021) Soil compaction due to increased machinery intensity in agricultural production: its main causes, effects and management. https://doi.org/10.5772/intechopen.98564. https://www.intechopen.com/chapters/77140#B24

Hakansson I, Reeder RC (1994) Subsoil compaction by vehicles with high axle load—extent, persistence, and crop response. Soil Tillage Res 29:277–304

Hamza M, Anderson W (2005) Soil compaction in cropping systems: a review of the nature, causes and possible solutions. Soil Tillage Res 82:121–145

Havlin JL, Beaton JD, Tisdale SL, Nelson WL. Soil fertility and fertilizers: an introduction to nutrient management, 6th Ed. 2005; Prentice Hall; Upper Saddle River, NJ

Horn R, Lebert M (1994) Soil compactability and compressibility. In: Soane BD, van Ouwerkerk C (eds) Soil compaction in crop production. Elsevier, Amsterdam, pp 45–69

Idowu J, Angadi S (2013). Understanding and managing soil compaction in agricultural fields. https://aces.nmsu.edu/pubs/_circulars/CR672.pdf

Jakobsen B, Dexter A (1987) Effect of soil structure on wheat root growth, water uptake and grain yield. A computer simulation model. Soil Tillage Res 10:331–345

Keller T, Sandin M, Colombi T, Horn R, Or D (2019) Historical increase in agricultural machinery weights enhanced soil stress levels and adversely affected soil functioning. Soil Tillage Res 194:104293

Kirby M, Bengough AG (2002) Influence of soil strength on root growth: experiments and analysis using a critical-state model. Eur J Soil Sci 53(1):119–127

Kumar V, Butter TS, Samanta A, Singh G, Kumar M, Dhotra B, Yadav NK, Choudhary RS (2018) Soil compaction and their management in farming systems: a review. Int J Chem Studies 6(3):2302–2313

Lipiec J, Stepniewski W (1995) Effects of soil compaction and tillage systems on uptake and losses of nutrients. Soil Tillage Res 35:37–52

Liu H, Colombi T, Jack O, Keller T, Weih M (2022) Effects of soil compaction on grain yield of wheat depend on weather conditions. Sci Total Environ 807:150763

Müller L, Lipiec J, Kornecki TS, Gebhardt S (2011) Trafficability and workability of soils. Encyclopedia Agrophys:912–924

Nawaz MF, Bourrie G, Trolard F (2013) Soil compaction impact and modeling: a review. Agron Sustain Dev 33:291–309

Nyeki A, Milics G, Kovacs AJ, Nemenyi M (2017) Effects of soil compaction on cereal yield: a review. Cereal Res Commun 45(1):1–22

Obour PB, Ugarte CM (2021) A meta-analysis of the impact of traffic-induced compaction on soil physical properties and grain yield. Soil Tillage Res 211:105019

Ojeniyi SO, Dexter AR (1979) Soil factors affecting the macrostructures produced by tillage. Trans Am Soc Agric Eng 22:339–343

Olubanjo OO, Yessoufou MA (2019) Effect of soil compaction on growth and nutrient uptake of Zea Mays L. Sustainable Agric Res 8(2):46–54

Parlak M, Parlak AO (2011) Effect of soil compaction on root growth and nutrient uptake of forage crops. J Food Agric Environ 9(3&4):275–278

Sivarajan S, Maharlooei M, Bajwa SG, Nowatzki J (2018) Impact of soil compaction due to wheel traffic on corn and soybean growth, development and yield. Soil Tillage Res 175:234–243

Soane BD, Van Ouwerkerk C (1998) Soil compaction: a global threat to sustainable land use. Adv GeoEcol 31:517–525

Taylor HM, Roberson GM, Parker JJ (1966) Soil strength-root penetration relations for medium- to coarse-textured soil materials. Soil Sci 102:18–22

Thomasson AJ (1982) Soil and climatic aspects of workability and trafficability. Proceedings of the 9th conference of the International Soil Tillage Research Organisation (ISTRO), Osijek, Yugoslavia, pp 551–557

Utomo WH, Dexter AR (1981) Soil friability. J Soil Sci 32:203–213

Voorhees WB (1986) The effect of soil compaction on crop yield. SAE Trans 95(3):1078–1084

Wang M, He D, Shen F, Huang J, Zhang R, Liu W, Zhu M, Zhou L, Wang L, Zhou Q (2019) Effects of soil compaction on plant growth, nutrient absorption, and root respiration in soybean seedlings. Environ Sci Pollut Res 26:22835–22845

Willatt ST (1986) Root growth of winter barley in a soil compacted by the passage of tractors. Soil Tillage Res 7:41–50

Chapter 9
Management of Soil Physical Environment in Relation to Plant Growth

Fluctuating climatic conditions including increasing uncertainties of rainfall events and droughts require adoption of management practices, which improve and sustain the health and resilience of agricultural systems. To sustain crop productivity, it is essential to sustain soil environment at a level optimum for plant growth. There are essentially three components of soil environment: soil physical, chemical, and biological environment. All the three components are strongly interdependent. Inadequacy of even one of them fails the agricultural production system to produce optimally. The basic soil physical properties directly affecting the plant growth and development are soil water, air, temperature, and mechanical impedance. Several technological options are available to modify these properties. Their choice, individually or in combination, depends on the agroecological situation. Some of these technologies are discussed in this chapter.

9.1 Tillage

The primary objective of tillage is to create seedbed having soil physical conditions favorable for seed germination, root growth, and crop development. Tillage is also practiced for mixing of manures and fertilizers in soil, incorporation of crop residues in soil, weed control, and soil aeration. Broadly speaking, there are two types of tillage systems: (i) conventional tillage and (ii) conservation tillage. In addition, there are two other tillage systems which are used under special conditions: (a) *soil compaction*, used in loose sandy soils, and (b) *wet tillage*, also called *puddling*, used in flooded rice soils.

P. K. Sharma, S. Kumar, *Soil Physical Environment and Plant Growth*,
https://doi.org/10.1007/978-3-031-28057-3_9

9.1.1 Conventional and Conservation Tillage

Conventional tillage is an intensive tillage system that includes primary (plowing and cutting of soil) and secondary tillage operations (harrowing and planking). It is a tillage system that has been developed and practiced traditionally by the farming communities. It is site, crop, and climate specific and also depends on the socioeconomic status of the farmers. The tillage equipment used vary widely in different regions of the world, ranging from primitive, light, animal-drawn countryside plows, and wooden planks (for land leveling, clod breaking, and pulverization) to heavy and sophisticated equipment like moldboard plows, disc plows, harrows, seeders, harvesters, tractors, and other farm traffic.

Conservation tillage is a system less intensive than conventional tillage, which aims at conserving resources like soil, labor, fuel, time, and cost. It refers to a tillage system in which at least 30% of the soil surface is covered by plant residues after planting (Conservation Technology Information Centre 1999). A minimum of 30% residue cover is needed to protect soil from water erosion or a minimum of 1.12 t/ha residue cover is needed to protect soil from wind erosion. However, the more the residue cover, the better it is. There are several variants of conservation tillage depending on the mode and the intensity of tillage operations, viz. no-till/zero-till/direct drilling system, slot planting, strip-till system, ridge-till system, stubble-mulch tillage, in-row subsoiling, reduced tillage, and minimum tillage.

Physical manipulation of soil through tillage operations modifies soil structure and structure-forming processes. Soil becomes loose, bulk density decreases, macroaggregates are broken down into micro-aggregates, and there is loss of soil organic carbon (SOC), especially the aggregate-bound carbon. Modification in soil structure triggers a change in associated soil physical properties and processes like soil strength, soil porosity, pore size distribution, soil water retention, infiltration rate, hydraulic conductivity, air exchange, and thermal properties. Tillage effects on some physical properties of soil are briefly discussed below:

- Soil aggregation

Conventional tillage, which is very frequent and intense, invariably results in reduction in macroaggregation and increase in micro-aggregation. Conservation tillage, on the other hand, which is less intensive and aims at conserving resources, conserves SOC and favors the formation of macroaggregates. The effect is more clear with no-till and minimum tillage systems compared to conventional till system (Wright and Hons, 2005; Abid and Lal, 2008; Zhang et al. 2012; Liu et al. 2021).

- Bulk density, soil strength, soil porosity, and soil-water-air relations

The impact of tillage on bulk density of soil varies with the tillage system, soil texture, cropping system, and climatic conditions. The effect of conventional tillage on bulk density is immediate while that of conservation tillage takes time to become visible. The immediate effect of conventional tillage is the loosening of surface soil and decrease in bulk density. Bulk density starts increasing as the crop growth progresses, and by the harvest stage, the bulk density may reach its initial value or may become even higher than its initial value. This effect is visible in mineral soils, which are low in organic matter content and where the soil has been intensively plowed with moldboard plow or disc plow. The conventional tillage over time often leads to subsurface compaction, especially if mechanized farming is practiced. The mechanical load of farm machinery and traffic (e.g., tractors, plows, harrows, seeders, harvesters) is responsible for subsoil compaction. Further, conventional tillage manipulates only the surface (0–15 cm) soil layer; the subsoil remains undisturbed.

Bulk density, in the long run, is generally lower under conservation tillage than under conventional tillage. Physical disturbance of soil under conservation tillage is either nil (as in zero-till or no-till system) or minimal (as under minimum tillage system), and there is buildup of SOC due to surface retention of organic residues. It improves soil aggregation, makes soil porous, and reduces bulk density. Different variants of conservation tillage, however, show different impacts on bulk density, and the responses further vary with soil texture, amount of crop residues returned to soil, cropping system, and climatic conditions (Ordoñez-Morales et al. 2019). In majority of cases, bulk density under no-till system is high compared to other conservation tillage systems (Unger and Jones, 1998; Abid and Lal, 2008).

The tillage system which changes bulk density also changes soil strength, total porosity, pore size distribution, and hydraulic properties of soil. Soil strength increases, while total porosity, infiltration rate, and saturated hydraulic conductivity decrease with the increase in bulk density. The volume of macropores decreases while those of micropores increases with the increase in bulk density (Douglas and Crawford, 1993). Ekeberg and Riley (1997) reported significant increase in the volume of macropores (>30 μm) in the top 30–70 mm soil layer with minimum tillage for 10 years. The continuity of soil pores is more under conservation than under conventional tillage because of higher earthworm activity and more root channels under conservation tillage system.

Increase in microporosity with the increase in bulk density increases soil-water retention below field capacity. It may also prolong the transient (second) stage of soil evaporation when evaporation rate is soil profile controlled (Sharma, 2017). Loose soil on the soil surface acts as dust mulch and reduces evaporation rate. Soil densification increases thermal conductivity of soils.

9.1.2 Wet Tillage

Wet tillage, also called *puddling*, is practiced for rice cultivation. Puddling is a process of repeated tillage under wet conditions with the objective of destroying soil aggregation and producing a soft plastic mud of low water permeability for ease of transplanting of rice and ensuring water stagnation in the field. Soil is converted from three-phase to two-phase system; the gaseous phase is practically eliminated. Chaudhary and Ghildyal (1969) reported a drop in mean weight diameter of aggregates from 1.70 mm to 0.36 mm due to puddling. It leads to reduction in macroporosity and increase in microporosity. Non-capillary porosity may reduce by

80–100% (Bodman and Rubin, 1948; Jamison, 1953). It severely affects soil water retention, hydraulic conductivity, and gaseous exchange.

Water retention at low water potentials is higher in puddled than in non-puddled soils. Puddled soils take longer time (weeks to months) to reach a workable moisture content as compared to well-aggregated soils because of low rates of evaporation (Gupta and Jaggi, 1979). Puddling decreases saturated hydraulic conductivity but increases unsaturated hydraulic conductivity. The gaseous exchange is severely restricted in puddled soils. It may lead to decline in O_2 concentration and buildup of CO_2 concentration in soil. Some soils may generate as high as 2.5 tons of CO_2 per hectare within 3 weeks of puddling and submergence (Ghildyal, 1982).

Puddling decreases bulk density and soil penetration resistance of surface layer, and the magnitude of reduction depends on puddling intensity, puddling equipment, soil texture, and initial aggregation status of soil. Bulk density immediately after puddling is low as compared to non-puddled soil, but it increases with time probably due to settling of soil particles and may reach a value which is higher even than its initial value (Table 9.1). Soil strength follows the trend of bulk density. Sharma and De Datta (1986) observed 0.1 and 0.5 MPa soil penetration resistances in 0–10 cm clay soil with and without puddling. Puddling in the long range often results in the formation of subsurface hardpans (Sharma and De Datta, 1986).

Puddling has significant effect on thermal properties of soil. In one study, puddling increased volumetric heat capacity and decreased thermal conductivity, thermal diffusivity, and damping depth (Sharma and De Datta, 1991). Sharma and De Datta (1988) observed maximum temperature of 28.1 °C in a puddled and submerged clay soil at 20 cm depth compared to 29.6 °C in non-puddled soil when the maximum air temperature 1 m above soil surface was 31.5 °C. The corresponding values in clay loam soil were 29.3, 32.4, and 32.5 °C, respectively. Time lag between maximum air temperature and soil temperature also increases with puddling and submergence. For example, time lag was 4.6, 6.4, 8.7, and 11.6 h at 50, 100, 150, and 300 mm depth in puddled and flooded soil compared to 2.0, 2.5, 3.0, and 6.0 h, respectively, in non-puddled soil (Chaudhary and Sandhu, 1982; Sharma and De Datta, 1991). Excellent reviews are available in literature on the effects of puddling on soil physical properties and processes (Sharma and De Datta, 1986; Kalita et al. 2020).

Table 9.1 Effect of puddling on the bulk density (Mg/m^3) of soil

	Puddled soil		Non-puddled soil		
Soil texture	At transplanting	At harvest	At transplanting	At harvest	Source
Clay	0.53	0.67	0.83	0.89	Sharma and De Datta (1986)
Clay loam	0.81	1.18	1.16	1.21	Sharma and De Datta (1986)
Silty clay loam	1.30	1.49	1.40	1.44	Bajpai and Tripathi (2000)

9.1.3 Soil Compaction

Generally, soil compaction is not desired in agricultural lands because it interferes with several soil and plant growth processes, which adversely affect the crop yields (Sects. 8.7 and 8.8). However, in some soils, compaction benefits crop growth and yield. Coarse-textured sandy soils in almost all ecosystems have high water permeability and low water retention. They also permit excessive nutrient leaching. Crops in these soils, especially in arid and semiarid regions, often suffer from water and nutrient stresses. Moderate compaction of highly permeable sandy soils improves soil-water retention and reduces leaching losses of nutrients. Compaction improves seed-soil contact and favors seed germination and uniform crop stand. In a field study, compaction of a loamy sand soil with the help of a 1000 kg iron roller increased bulk density of 0–15 cm soil layer on average from 1.48 to 1.57 g/cc, decreased infiltration rate from 35.2 to 21.6 cm/h, decreased saturated hydraulic conductivity from 10.85 to 8.16 cm/h, increased soil moisture content from 8.1 to 13.1%, and significantly increased yield of various crops including pearl millet, guar, cowpea, wheat, barley, raya, taramira, fenugreek, tomato, and chilies (Majumdar, 1994; Majumdar et al. 2000). In another study (Sharma et al. 1995), subsoil compaction improved soil-water relation in a loamy sand soil, which resulted in significant increase in grain yield and water-use efficiency of rainfed lowland rice (Table 9.2). Subsoil compaction was achieved with 6–9 passes of a 12-t road roller followed by disc plowing of top 20 cm soil layer.

Table 9.2 Effect of subsoil compaction on percolation, water retention, grain yield, and water-use efficiency of rainfed lowland rice in a loamy sand soil (Adapted from Sharma et al. 1995)

Parameter	Dry tillage (non-compacted)[a]	C9-V (compacted)[b]
Percolation rate (mm/d)	11.8	1.4
Drought days (no.)[c]	63	6
Soil moisture at rice harvest (mm/30 cm)	27.1	42.8
Soil moisture 20 days after rice harvest (mm/30 cm)	21.3	32.6
Grain yield (Mg/ha): 1992ws	2.4	4.0
Grain yield (Mg/ha): 1993ws	0.9	2.1
Water-use efficiency (kg/ha-mm):		
1992 ws	3.10	5.06
Water-use efficiency (kg/ha-mm):		
1993 ws	1.80	4.34

ws Wet season

Note: Means for all parameters between compacted and non-compacted treatments are significantly different ($P = 0.05$)

[a]Soil was dry-tilled with disc plow to 15 cm depth followed by submergence with rainwater

[b]Soil was compacted with 9-passes of 12-t road roller with vibration-on mode

[c]Number of days the rice fields remained without water on the soil surface

9.2 Soil Amendments

Soil amendments are the materials incorporated into (problem/degraded) soils to improve their physical, chemical, and biological properties in relation to plant growth and yield. The primary objective of use of soil amendments in agriculture is to make soils productive. Soil amendments perform different functions, such as:

- Improvement of soil physical properties, viz. soil texture, soil structure, soil-water-air relations, soil thermal regimes, and drainage conditions
- Improvement of soil chemical properties, viz. soil pH, nutrient status, and soil salinity
- Improvement of biological properties of soil, viz. microbial population and their activities
- Reduction of susceptibility of soils to runoff and erosion
- Improvement of degraded lands and protection of lands from further degradation

Soil amendments have long been used to ameliorate degraded lands, viz. salt-affected soils, acid sulfate soils, waterlogged soils, eroded soils, and sandy soils. Soil amendments may be of organic or inorganic origin. They may be natural or synthetic materials. The natural organic soil amendments are of plant and animal origin, viz. animal dung (cow or chicken), green manure, compost, vermicompost, peat and moss, wood chips, and biochar. The natural inorganic soil amendments are derived from mining, such as lime, gypsum, and sulfur. Synthetic organic amendments include *soil conditioners*, which are long-chain organic polymers used primarily to improve soil structure. The choice of amendment(s) depends primarily on the nature of soil degradation and the cost involved in soil amendment. Different types of soil amendments along with their salient functions are summarized in Table 9.3.

- *Organic manures*: The *animal dung* (farmyard manure) and *compost* are most often easily available, economically affordable, and hence most commonly used organic amendments worldwide. They have two limitations of having offensive odor and bulkiness. *Green manure* is also a common soil amendment, next to organic manures (FYM, compost) in popularity. Raising a green manure crop requires 6–8-week period and availability of inputs like seed, irrigation, and nutrient sources. Green manure crops should be of short duration, quick growing, producing large biomass, and capable of growing in poor marginal lands. Leguminous crops are usually preferred as green manure crops. *Peat and moss* are mostly used in commercial agriculture. All organic amendments improve SOC content, which leads to increase in soil aggregation, soil porosity, soil water retention, and decrease in soil bulk density and soil strength, with the overall improvement in soil physical health (Cercioglu, 2017).
- *Biochar*: Biochar is seen as a potential soil amendment. It has the potential for carbon sequestration in soils because of its highly recalcitrant carbon content and improves soil aggregation (macroaggregates), aggregate stability, and related soil properties (Lehmann and Joseph, 2009; Mukherjee and Lal, 2016; Widowati

Table 9.3 Soil amendments and their salient functions

Soil amendments		Salient functions
Natural amendments	*Organic*	
	Farmyard manure, compost, vermicompost, green manures, crop residues, rice husk, coconut fiber, sawdust, woodchips, etc.	Add organic matter to soil; improve soil structure (aggregation, porosity, bulk density) and soil-air-water relations; reduce susceptibility to runoff and erosion; improve biological activities in soil; supply essential plant nutrients and hormones, etc.
	Peat and moss	Increases soil-water retention (ideal for sandy or rocky soils); stabilizes clay soils
	Biochar	Adds organic carbon to soil; improves soil physical conditions
	Inorganic	
	Lime	Increases pH of acidic soils; improves soil structure and related soil physical properties, etc.
	Gypsum, phosphogypsum	Replaces Na with Ca ions to improve sodic soils (SAR > 15); improves saline soils in irrigated areas; improves soil structure and related soil physical properties, etc.
	Sulfur	Lowers soil pH; improves saline soils
	Clay (bentonite, vermiculite, perlite[a]), silt	Improves soil texture and increases soil water retention of sandy soils
	Sand	Improves soil texture, drainage, and aeration conditions of clay soils
Synthetic amendments	*Organic*	
	Synthetic soil conditioners	Improve soil structure and soil-water relations; reduce susceptibility of soils to wind erosion, etc.

[a]Made from heated amorphous volcanic glass

et al. 2020; Sun et al. 2021). It has the advantage of being free from offensive odor and bulkiness, unlike fresh farmyard manure and composts. The properties of biochar as a soil amendment depend on the pyrolytic process (pyrolysis temperature and charring time) and the nature of the substrate from which biochar is prepared (organic wastes, such as livestock manures, sewage sludge, crop residues, composts, weeds, wood chips, etc.). Studies indicate more positive effects of biochar on soil water retention in coarse- to medium-textured soils than in fine-textured soils (Jeffery et al. 2011). Application rate of 0.25–2% (g/g) biochar can significantly improve soil physical health in terms of water-stable aggregates and soil water retention.

- *Inorganic soil amendments*: Lime and gypsum are commonly used inorganic amendments to ameliorate acidic and salt-affected soils, respectively. These soils have low chemical fertility such as deficiency/toxicity of macro- and micronutrients. Acid soils exhibit Al, Fe, and Mn toxicity and deficiency of Mg, Ca, P, and Mo. The salt-affected soils exhibit osmotic stress, oxidative stress, ion toxicity (Na, B, Cl), nutrient (N, Ca, K, P, Zn) deficiency, etc. The physical properties of these soils are also inadequate for plant growth. Acidic soils are more vulnerable to soil structure decline and erosion. Application of lime or gypsum/phosphogypsum improves soil structure (i.e., soil aggregation) and corrects nutritional disorders (toxicity/deficiency) by way of correcting (increasing/decreasing) soil pH and exchanging H^+ and Na^+ ions on clay particles by Ca^{2+} and Mg^{2+} ions.

- *Synthetic soil conditioners*: Use of synthetic soil conditioners as soil amendments is fairly new. These products are promoted as "soil conditioners." *Soil conditioners* are synthetic organic long-chain polymers used primarily to improve soil physical properties. They behave like polycations [dimethylaminoethyl methacrylate (DAFMA)] or polyanions [polyvinyl acetate (PVA_c), polyacrylonitrile—partly hydrolyzed (H_pPAN) or hydrolyzed (HPAN)—polyacrylic acid (PAA), vinyl acetate-maleic acid copolymer (VAMA)] or dipole polymers [polyacrylamide (PAM)] or non-ionized polymers [polyvinyl alcohol (PVA)] or emulsions (bitumen). Several new products with different brand names are now available in the market. They bind soil particles into aggregates by forming ionic bridges between them or by coating the inert sand and silt particles like glue (Deboodt, 1972). They are either mixed with soil as dry powder or sprayed on the soil surface (with or without mixing with soil). The choice of a soil conditioner depends on the soil, climate and crop conditions, and cost of application. They have been successfully used in stabilizing sand dunes and road embankments, checking wind erosion, etc. But their use in agriculture has remained limited because of high cost, restricted availability, and application difficulties. However, they are now receiving increased attention as potential erosion control practices, and efforts are being made to make them economically feasible for their agricultural use. Several studies have shown significant impact of soil conditioners on soil physical, chemical, and biological properties of soil (Sharma, 1988a,b; Steinberger et al. 1993; Hayat and Ali, 2004; Alkhasha et al. 2018; Zein El-Abdeen, 2018).
- *Clay as a soil amendment*: Mixing of fine-textured material (clay and silt) with the coarse-textured sandy soils changes the soil texture, decreases water permeability, and increases soil water retention. The process is known as *soil dressing*. The fine-textured soil may be transported from the nearby areas, if available, and mixed with the surface sandy soil. The technique is expensive but has long-lasting effects. If the sandy soil is underlain by a soil layer rich in fine-textured material, within the reach of tillage equipment, deep tillage can be used to mix the fine-textured material underneath and the coarse-textured material on the surface. The depth of profile modification (C_d) needed to obtain specified clay content in the modified soil layer ($C_{\%c}$) depends on the clay contents of surface and subsurface soil layers.

The depth of profile modification may be determined by using the following equation provided that the initial clay contents of the surface coarse layer ($A_{\%c}$) and subsurface fine layer ($B_{\%c}$) and the depths of surface layer (A_d) and subsurface layer (B_d) are known (Unger et al. 1981):

$$\left(A_d \times A_{\%c}\right) + \left(B_d \times B_{\%c}\right) = \left(C_d \times C_{\%c}\right) \tag{9.1}$$

$$\text{and} C_d = A_d + B_d \tag{9.2}$$

Substituting the value of C_d from Eq. 9.2 in Eq. 9.1 and rearranging, the equation becomes

$$B_d = \left\{A_d\left(C_{\%c} - A_{\%c}\right)\right\} / \left(B_{\%c} - C_{\%c}\right) \tag{9.3}$$

The solution of Eq. 9.3 gives the value of depth of subsoil layer (B_d) to be taken into consideration for deep tillage. Using this value in Eq. 9.2, the depth of modified layer (C_d) can be calculated; the depth of surface layer (A_d) to be modified is prefixed.

9.3 Cropping Systems

Cropping systems and the associated tillage practices significantly affect soil physical properties (Pikul Jr et al. 2006). Cover crops, heavy-residue crops, deep-rooted crops, sod-based rotations, and legume crops in cropping systems improve soil physical properties, and the effect is multiplied if such cropping systems are coupled with conservation tillage. Cover crops improve soil physical properties and decrease susceptibility of soils to erosion. Deep-rooted crops may puncture compact subsoil layers, increase water permeability, and recycle plant nutrients from deeper layers to surface soil layer. Heavy-residue crops return increased amount of biomass to soil, thereby improving SOC content and biological activity in soils.

Although contrasting findings have been reported in literature, majority of studies reveal positive effects of *cover crops* on soil physical properties. Permanent cover crops are more effective than the annual cover crops rotated with other field crops. Irmak et al. (2018) did not observe any significant impact of cover crops grown for 12 years on the physical properties of a silt loam soil. Blanco-Canqui and Ruis (2020), on the other hand, reviewed 98 field studies and concluded that in a majority of cases cover crops on average reduced soil penetration resistance by 5%, improved wet aggregate stability by 16%, and increased cumulative infiltration by 43%, but showed negligible effect on bulk density, dry aggregate stability, saturated and unsaturated hydraulic conductivity, field capacity, and plant available water capacity. Soils under cover crops were generally cooler in daytime and warmer at nighttime, and warmer in winter and cooler in the rest of the year. Cover crops moderated thermal conductance of soils by increasing their volumetric heat capacity and decreasing thermal diffusivity. The impact of cover crops on soil physical properties varied with soil texture, tillage system, and duration and SOC contribution of cover crops. The positive effect of cover crops on soil physical properties increased with the duration of cover crops coupled with no-till system. With the thinking that permanent cover crops improve soil physical health, the USDA initiated the Conservation Reserve Program (CRP) in 1986 to stop crop cultivation on erosion-susceptible lands and put them under permanent vegetation cover (Wu and Xie, 2017).

Long-term use of intensive cropping systems coupled with conservation tillage has the potential of C and N sequestration leading to improved soil physical properties in relation to plant growth (Peterson et al. 2002; Wright and Hons, 2005). In a 20-year field study in south-central Texas (USA), the SOC and SON sequestration in 0–5 cm soil was significantly greater under no-till (NT) than conventional till (CT) system for a grain sorghum-wheat-soybean SWS) rotation, a wheat-soybean double crop (WS), and a continuous wheat monoculture (CW) (Wright and Hons, 2005). NT increased SOC storage compared to CT by 62%, 41%, and 47% and SON storage by 77%, 57%, and 56%, respectively, for SWS, WS, and CW cropping sequences. The increase in SOC content leads to increased soil aggregation and aggregate stability and decreased bulk density and soil strength (Zhou et al. 2020).

The inclusion of upland crops in lowland rice culture significantly improves physical properties of lowland rice soils. Ba et al. (2016) reported higher SOC, lower bulk density and soil strength, and higher total porosity, aggregate stability index, and plant available water capacity in a clay rice soil under rice-maize-rice, rice-mung bean-rice, and rice-mung bean-maize cropping than under rice-rice-rice cropping system.

9.4 Mulching

Any material spread on soil surface as a covering is known as *mulch*, and the practice of application of mulch is known as *mulching*. Mulch acts as a barrier between soil and outer atmosphere. Different types of materials, organic and inorganic, can be used as mulch, viz. *organic mulches* (crop residues, weed biomass, leaves, wood chips, forest litter, FYM, compost, saw dust, cocopeat, etc.) and *inorganic mulches* (plastic sheets (transparent, opaque, black), gravels, geotextiles, etc.). The choice depends on the objective of mulching and cost and availability of mulch material.

Mulch farming is a form of conservation tillage that preserves soil quality and the environment. Mulching affects physical, chemical, and biological properties of soil by increasing SOC content, acting as a physical barrier between soil and outer atmosphere, and protecting surface soil against the beating action of raindrops, runoff, and high wind velocities. The effect depends on the nature, quality, and quantity of the mulch material used.

Any decomposable organic material that can be used to cover the soil surface is organic mulch. Organic mulches, in addition to acting as a physical barrier on the soil surface, have added advantage of adding organic carbon (humus) to soil upon decomposition. The SOC buildup significantly improves soil structure and related soil physical properties (porosity, aggregation, aggregate stability, bulk density, soil strength, soil's water and air permeability, and water retention capacity). Mulching reduces evaporation, conserves soil moisture, and regulates soil temperature. Several plant nutrients are released in soil during the decomposition of organic mulches. Organic mulches also activate soil microflora and enhance the population of beneficial soil microorganisms. Organic mulches, thus, improve the overall quality of soil. Plastic mulches regulate soil moisture, evaporation, and soil thermal regimes, but practically have no effect on soil structure.

Mulching reduces evaporation from the soil surface, especially during the initial constant rate stage when evaporation rate is controlled by atmospheric evaporativity (Sharma, 2017). Mulching reflects a part of solar radiation and reduces its intensity on the soil surface. It also decreases wind velocity on the evaporating soil surface by creating a physical barrier. Low solar radiation and low wind velocity at soil surface decrease atmospheric evaporativity and, hence, rate of evaporation from the soil surface. Water gets time to move into deeper soil layers through internal distribution, thereby improving soil water conservation. Mulching prolongs the first stage (constant rate stage) of evaporation and keeps the surface soil moist for a longer period of time. Evaporation is affected more by plastic sheet mulches than organic mulches.

Mulching is very effective against soil erosion. Mulch intercepts rainwater and protects soil aggregates from breaking. It prevents soil crusting, avoids surface sealing, and increases or maintains infiltration rate. In addition, physical resistance offered by mulch material to water flow reduces the amount and velocity of runoff. It protects soil from water erosion. Mulch protects soil from wind erosion as well as by reducing wind speed on the soil surface.

Mulch affects soil thermal regime by modifying the intensity of incident solar radiation on the soil surface and the thermal properties of soil. Reduction in radiation intensity, bulk density, and thermal conductivity and increase in volumetric heat capacity of soil have a moderating effect on soil thermal regime. The effect varies with the nature and amount of mulch material, soil type, and climatic conditions. Soils under mulch are warmer during winters and cooler during summers than the soils without mulch. Mulch reduces diurnal temperature fluctuations in soils. Nighttime temperature is more, while daytime temperature is lesser in soils under mulch than without mulch. Organic mulch (e.g., straw mulch) lowers the maximum soil temperature, transparent plastic sheet mulch raises it, while black polyethylene sheet has nil to negligible effect on maximum soil temperature.

9.5 Nutrient Management

Nutrient requirement of crop plants is met through soil and externally applied nutrients as inorganic fertilizers and organic manures. Nutrient stock in soil is limited and gets progressively exhausted under continuous intensive cropping if not supplemented externally. Intensive cropping, especially of high-yielding cultivars, has increased nutrient dependence of crop plants on external sources. Several long-term field studies have shown that intensive use of chemical fertilizers significantly affects soil physical properties. The effect may be positive or negative depending on whether chemical fertilizers are used in balanced or imbalanced form and applied along with or without organic manures. The effect also varies with soil texture and type and amount of fertilizers used.

Use of chemical fertilizers at recommended rates generally sustains soil health and crop productivity. Studies also report improvement in soil physical properties. Adequate and balanced fertilizer use supports a healthy crop growth with increased biomass production, increased return of organic matter to soil through root biomass and crop residue litter, enhanced biological activities in soil, and improvement in soil physical properties (Manna et al. 2005). Integration of chemical fertilizers with organic sources (FYM, compost, green manure, vermicompost, crop residues, etc.) has an additive effect (Verma, 2006; Saha et al. 2010; Nandapure et al. 2011; Brar et al. 2015; Dhaliwal et al. 2019). There is significant improvement in soil microbial mass and enzymatic activities (Liu et al. 2010), which have a positive effect on soil properties and productivity. The long-term effect of integrated nutrient management on some soil physical properties in an acidic clay loam soil is shown in Table 9.4.

The chemical fertilization regime, especially in imbalanced form, is unsustainable and detrimental to soil health. There is overall decline in soil chemical, biological, and physical properties. Long-term use of nitrogenous fertilizers alone (e.g., urea, ammonium sulfate, and ammonium nitrate) lowers soil pH, reduces microbial population (bacteria, fungi, and actinomycetes), and decreases important enzymatic activities (dehydrogenase, alkaline phosphatase, β-glucosidase, urease) (Liu et al. 2010). Use of only N fertilizers often results in deficiency of other nutrients (e.g., P and K) in soil. Use of N alone may damage soil structure and related soil physical properties. Conditions unfavorable for nitrification (e.g., low pH, high levels of

Table 9.4 Long-term effects of different nutrient management practices on some physical properties of a silt loam soil under maize-wheat cropping (1972–1973 to 2003–2004) (adapted from Verma, 2006)

Treatment	Bulk density (g/cc)	MWD (mm)	WSA (>0.25 mm (%)	Total porosity (%)	Aggregate porosity (2–8 mm) (%)	IR ($\times 10^{-6}$ m/s)	Ks ($\times 10^{-6}$ m/s)	PAWC (% by wt)
Control	1.34	1.36	71.0	47.8	38.5	3.65	4.59	12.1
N	1.35	0.97	62.9	47.5	34.2	3.49	3.99	11.6
NP	1.23	1.59	77.5	52.0	42.0	5.47	9.96	14.0
NPK	1.22	1.67	78.6	52.5	43.2	5.70	11.14	14.2
NPK+ FYM	1.15	2.52	83.6	54.9	51.7	7.44	18.57	16.6
NPK+ Lime	1.32	3.59	89.6	48.3	39.6	3.81	6.98	13.0
CD (0.05)	0.04	0.22	3.7	1.5	2.1	0.48	2.19	0.7

Note: 1. Control: no fertilizer/FYM/lime applied
2. Soil sampling depth = 0–15 cm

Table 9.5 Long-term effects of different nutrient management practices on particle size distribution in a silt loam soil under maize-wheat cropping (1972–1973 to 2003–2004) (Verma, 2006)

	Sand		Silt		Clay	
Treatment	0–15 cm	15–30 cm	0–15 cm	15–30 cm	0–15 cm	15–30 cm
Control	23.5	24.0	53.4	49.3	23.2	26.7
N	21.6	22.7	52.8	50.0	25.5	27.3
NP	22.5	24.7	53.8	48.7	23.7	26.7
NPK	22.7	25.3	53.7	49.3	23.6	25.3
NPK + FYM	22.5	22.3	53.7	48.7	23.8	26.0
NPK + lime	23.2	25.3	53.4	49.3	23.4	25.3
CD (0.05)	1.0	ns	ns	ns	0.6	ns

Fertilizer application rates:
Maize: 120 kg N-60 kg P_2O_5–40 kg K_2O ha^{-1}
Wheat: 120 kg N-60 kg P_2O_5–30 kg K_2O ha^{-1}
FYM and lime ($CaCO_3$) applied to maize crop only @ 10 t/ha and 0.9 t/ha, respectively

accumulated NH_4^+, low soil moisture content), created by long-term use of only N fertilizers, may lead to accumulation of NH_4^+ ions in soil and, like Na^+ ions, cause dispersion of soil colloids and aggregates. Excessive use of urea reduces soil pH and may affect even the soil texture. Under extreme acidic conditions, the dissolution of iron may occur, a process called *ferrolysis*, which disintegrates mineral fraction of soil. In a long-term field experiment in a silt loam soil under maize-wheat cropping, continuous application of urea (without P, K, and FYM) for 32 years decreased soil pH from the initial value of 5.8–4.4 and increased clay content of top 12 cm soil layer by 1.7–3.0% with a corresponding decline in sand fraction (Table 9.5). The textural class of 4–10 cm soil layer changed from silt loam to silty clay loam. Long-term application of P fertilizer in excess of crop requirement results in the buildup of large amounts of P in soil in both the inorganic and organic pools (Singh et al. 2007).

9.6 Livestock Grazing

Livestock grazing is a method of animal husbandry wherein domestic livestock are allowed to roam about freely outdoor to consume forages. Livestock grazing has a significant impact on soil physical properties. It may accelerate soil erosion and land degradation. Livestock grazing reduces land cover and causes degradation of soil physical properties. The trampling of ground by livestock may cause soil compaction, increase in bulk density, loss of aggregation, change in pore volume and function, decrease in macroporosity, and permeability and infiltration (Donovan and Monaghan, 2021). The relationship between ground cover and surface erosion is significant and nonlinear. Soil loss increases rapidly as the soil cover falls below 30% (Greene et al. 1994; Silburn et al. 2011).

The impact of grazing on soil physical properties depends on the grazing intensity which is affected by stocking density, hoof pressure, grazing duration, and history of previous grazing. For a given grazing intensity, soil degradation depends on soil moisture and clay content. Livestock trampling is more detrimental on clay and wet soils than sandy and dry soils. Reduction in soil porosity, especially the macroporosity, reduces water permeability and increases overland flow of water, thereby increasing the surface soil erosion potential.

The Universal Soil Loss Equation (USLE) has largely been used at local, regional, and national scales to estimate mean annual soil loss through surface erosion. It incorporates the following five factors to estimates soil loss: soil erodibility (K), ground cover (C), rainfall erosivity (R), slope length (L), and steepness of slope (S). It does not, however, consider the impact of livestock grazing on land cover and soil physical properties. Studies have shown that inclusion of these two parameters as a function of grazing intensity (density, duration, history, and stock type) as well as soil clay and water contents in USLE improves the prediction of soil losses through erosion (Donovan and Monaghan, 2021).

Nearly 40% of earth's ice-free terrestrial surface is covered by grasslands, and the forages are the largest consumed livestock feed in the world. Since irrational livestock grazing leads to deterioration in soil physical properties and soil health, the grazing management becomes crucial to avoid degradation of grasslands and to sustain forage-livestock production systems. *Grazing management* refers to "the manipulation of grazing in pursuit of a specific objective or a set of objectives" (Allen et al. 2011). The key components of grazing management are grazing intensity (i.e., stocking rate), grazing frequency (i.e., stocking method), and timing of grazing. The *grazing intensity* depends on the carrying capacity of the pasturelands. It considers the stocking rate, i.e., livestock units or livestock weight per unit area, and the quantity of forage available per unit area. The *grazing frequency* considers the allocation of animals to the pastures during grazing season and the choice of stocking methods. The grazing may be continuous, seasonal, or rotational within a grazing period. The *timing of grazing* considers the plant growth stage or season of the year. In case of annual grasses or short-lived grass species, which rely on natural reseeding for their regeneration, the timing of termination of grazing becomes very important. It depends on the timing of flowering and seed set of grass species. For details on grazing management, one may refer Sollenberger et al. (2020).

Adaptive Multi-Paddock (AMP) Grazing has been found better than continuous grazing with light or heavy stocking. In AMP system of grazing, small paddocks are grazed with high densities of livestock for short periods, keeping long recovery periods prior to re-grazing. One paddock is grazed at a time while other paddocks are allowed to recover, and the livestock number is adjusted according to change, if any, in the available forage conditions. The AMP grazing favours more grass production and increased carbon sequestration in soil. In one study in north-central Texas (USA), AMP grazing sequestered an additional 12 t of C/acre over a 10-year period as compared to other conventional grazing systems (e.g. continuous grazing with light or heavy stocking) (Teague et al. 2016).

9.7 Question Bank

9.7.1 *Short Questions:*

(i) What is tillage? What are the objectives of tillage? Distinguish between conventional and conservation tillage.
(ii) How does tillage affect soil physical properties?
(iii) What is puddling? How does it affect soil physical properties?
(iv) What is a soil amendment? Why are they used? Enlist different types of soil amendments.
(v) What is biochar? What is its impact on soil properties?
(vi) Write short notes on:

  (a) Biochar
  (b) Inorganic soil amendments
  (c) Synthetic soil conditioners
  (d) Clay as a soil amendment

(vii) What is the impact of cropping systems on soil physical properties? What type of crops needs to be included in cropping systems intended to improve soil physical properties?
(viii) Define *mulch* and *mulching*. What is the role of mulching on soil physical properties?
(ix) What is *integrated nutrient management* (INM)? Compare and contrast the effect of chemical fertilization regime and INM on soil physical properties and soil productivity.
(x) What is the impact of livestock grazing on soil physical properties of pasturelands? Briefly discuss important factors affecting the impact of livestock grazing on soil properties.
(xi) What are the key elements of livestock grazing management?

9.7.2 *Briefly Explain Why?*

(i) Conservation tillage improves and sustains soil physical properties.
(ii) Conventional tillage generally destroys soil structure and related soil physical properties.
(iii) Intensive tillage operations destroy soil aggregation.
(iv) Puddling destroys soil aggregates.
(v) Puddled soils are two-phase system.
(vi) Time lag between maximum air temperature and soil temperature also increases with puddling and submergence.
(vii) Soil compaction is beneficial for crop production in some soils.

(viii) Green manure is not as common with farmers as organic manures as soil amendment.
(ix) Biochar has the potential of carbon sequestration in soil.
(x) Soil conditioners in spite of their encouraging results in modifying soil physical properties have not become very popular in the agriculture sector.
(xi) Clay also serves as a good soil amendment.
(xii) The nature of mulch material is important in modifying soil physical properties.
(xiii) Organic mulches should be preferred over plastic mulches to improve soil structure.
(xiv) Cover crops, deep-rooted crops, heavy-residue crops, and leguminous crops improve soil physical properties.
(xv) The effect of mulch on reducing evaporation is more during the first stage of evaporation, i.e., during the initial constant rate stage as compared to second or third stage of evaporation.
(xvi) Mulching has a strong influence on soil thermal regime.
(xvii) INM is a better approach than chemical fertilization regime in order to sustain soil productivity.
(xviii) Use of N fertilizers alone is unsustainable and detrimental to soil health.
(xix) Livestock grazing has a significant impact on soil physical properties, soil erosion, and soil degradation of pasturelands.
(xx) The impact of livestock grazing on soil physical properties varies with soil texture and moisture content.
(xxi) Grazing intensity is the most important element of grazing management.

9.7.3 *Fill in the Blanks:*

(i) Broadly speaking, there are two types of tillage systems: (a) .... and (b) .... tillage system.
(ii) .... is an intensive tillage system.
(iii) Leaving at least ....% crop residues on the soil surface at crop harvest is an essential feature of conservation tillage.
(iv) A minimum of ......t/ha residue cover is needed to protect soil from wind erosion.
(v) Minimum tillage is one of the variants of .... tillage system.
(vi) Intensive tillage operations generally destroy soil aggregation by (a) .... and (b) ....
(vii) Wet tillage is popularly known as ....
(viii) Moderate compaction of sandy soils is desirable in crop cultivation as it enhances .... and decreases ....
(ix) .... are the materials incorporated into soils to improve properties and productivity of problem or degraded soils.
(x) The synthetic organic polymers used as soil amendments are called ....

(xi) ………… is used to ameliorate acidic soils.
(xii) ………… is used to ameliorate saline soils.
(xiii) …………… is an example of soil conditioner which behaves like a dipolymer in soil.
(xiv) ………… is an example of emulsion type of soil conditioner.
(xv) The mixing of fine-textured soil with sandy surface soil is known as ……….
(xvi) The process of spreading any decomposable organic material on soil surface as a soil cover is known as …………
(xvii) …………… mulch practically has no effect on soil structure.
(xviii) Organic mulches improve soil physical properties by adding ……….. to soil.
(xix) Mulching affects incident radiation on the soil surface by changing ……… on the soil surface.
(xx) Mulch ……… diurnal temperature fluctuations in soils.
(xxi) The chemical fertilizer regime, especially in imbalanced form, ……….. soil productivity in the long run.
(xxii) The long-term use of only N fertilizers often results in the deficiency of …… and …… in soil.
(xxiii) The long-term use of only N fertilizers may lead to accumulation of ……… ions in soil, which may cause dispersion of soil colloids and aggregates.
(xxiv) The key components of grazing management are (a) ………, (b) ………..., and (c) …………….
(xxv) Unscientific livestock grazing reduces land cover and exposes land to ……….

[**Key:** i. conventional, conservation; ii. Conventional tillage; iii. 30; iv. 1.12; v. conservation; vi. physical destruction, reducing SOC; vii. puddling; viii. seed-soil contact or available soil water capacity, nutrient leaching; ix. Soil amendments; x. soil conditioners; xi. Lime; xii. Gypsum; xiii. Polyacrylamide (PAM); xiv. Bitumen; xv. soil dressing; xvi. mulching; xvii. Plastic sheet; xviii. humus or organic carbon; xix. albedo; xx. reduces; xxi. decreases or spoils; xxii. P, K; xxiii. NH_4^+; xxiv. grazing intensity, grazing frequency, timing of grazing; xxv. soil erosion]

9.7.4 State Whether the Following Statements Are True (T) or False (F):

(i) Weed control is the primary objective of tillage.
(ii) Minimum tillage is a variant of conservation tillage.
(iii) Moderate soil compaction is beneficial to crops in sandy soils.
(iv) In conservation tillage, the entire soil surface is kept covered with crop residues at harvest.
(v) Tillage is an important tool to modify soil physical properties.
(vi) Conventional tillage over time often leads to subsurface compaction.

(vii) Bulk density under conservation tillage is comparatively higher than conventional tillage because of less or no disturbance of surface soil.
(viii) Volume of macropores is higher under conservation than under conventional tillage.
(ix) Soil compaction increases soil's thermal conductivity.
(x) Puddling converts soil from three-phase to two-phase system.
(xi) Porosity of puddled soils may be as low as zero.
(xii) Puddled soils take more time to dry as compared to dry-tilled soils.
(xiii) Time lag between occurrence of peak temperatures on the soil surface and deeper layers reduces with wet tillage.
(xiv) Moderate soil compaction of sandy soils improves seed germination and seedling emergence by increasing seed-soil contact.
(xv) Improvement of degraded soils with the use of soil amendments is an age-old practice.
(xvi) Soil conditioners have been used since long to improve soil structure.
(xvii) The synthetic organic long-chain polymers are called soil conditioners.
(xviii) Biochar has the potential of carbon sequestration in soils because of its highly recalcitrant carbon content.
(xix) Offensive odor and bulkiness are two problems of organic manures.
(xx) Biochar has two problems of having offensive odor and bulkiness in handling.
(xxi) Soil conditioners in soil behave like polycations or polyanions or dipole polymers.
(xxii) Clay can be used as a soil amendment in sandy soils.
(xxiii) Cover crops having deep-root system and heavy residues are best for improving soil physical properties.
(xxiv) Cropping systems have a significant effect on soil physical properties.
(xxv) Cropping systems including cover crops and coupled with conventional tillage significantly improve soil physical properties.
(xxvi) Mulching has a significant effect on soil physical environment.
(xxvii) Plastic mulches are more effective than organic mulches in improving soil structure.
(xxviii) Organic mulch is more effective in modifying soil thermal regime than plastic sheet mulch.
(xxix) Mulch reduces diurnal temperature fluctuations in soils.
(xxx) Chemical fertilization regime is not always destructive to soil physical properties.
(xxxi) Long-term use of only N fertilizers is detrimental to soil health.
(xxxii) Long-term use of only N fertilizers may lead to accumulation of NH_4^+ ions in soil, which helps in improving soil aggregation.
(xxxiii) Unscientific livestock grazing may expose pasturelands to increased soil erosion.
(xxxiv) Livestock trampling is more detrimental on sandy and dry soils than clay and wet soils.
(xxxv) The grazing intensity depends on the carrying capacity of the pasturelands.

[**Key:** i. F, ii. T, iii. T, iv. F, v. T, vi. T, vii. F, viii. T, ix. T, x. T, xi. F, xii. T, xiii. F, xiv. T, xv. T, xvi. F, xvii. T, xviii. T, xix. T, xx. F, xxi. T, xxii. T, xxiii. T, xxiv. T, xxv. F, xxvi. T, xxvii. F, xxviii. F, xxix. T, xxx. T, xxxi. T, xxxii. F, xxxiii. T, xxxiv. F, xxxv. T]

9.7.5 Multiple-Choice Questions:

(i) The primary objective of tillage is

(a) Weed control
(b) Mixing of fertilizers in soil
(c) Mixing of residues in soil
(d) Preparation of good-quality seedbed

(ii) Soil compaction is

(a) Always detrimental to crop production
(b) Good for crop cultivation in clayey soils
(c) Good for crop cultivation in sandy soils
(d) Good for cultivation in organic soils

(iii) Wet tillage

(a) Coverts soil from two-phase to three-phase system
(b) Reduces porosity to as low as 0%
(c) Lowers bulk density of surface soil layer
(d) Increases soil penetration resistance

(iv) Soil compaction

(a) Decreases microporosity
(b) Decreases macroporosity
(c) Decreases soil penetration resistance
(d) Decreases bulk density of soil

(v) The essential requirement of conservation tillage is to retain

(a) At least 30% crop residues on soil surface
(b) At least 50% crop residues on soil surface
(c) At least 70% crop residues on soil surface
(d) Al least 90% crop residues on soil surface

(vi) To protect soil from wind erosion, the soil surface must have residue cover of at least

(a) 1.12 t/ha
(b) 2.12 t/ha
(c) 3.12 t/ha
(d) 4.12 t/ha

(vii) Conservation tillage helps in improving soil structure primarily by

(a) Avoiding physical disturbance of soil
(b) Adding organic carbon/humus to soil
(c) Protecting soil from beating action of raindrops
(d) Avoiding abrasive action of wind

(viii) Any tillage practice which conserves SOC

(a) Improves microporosity
(b) Improves soil aggregation
(c) Suppresses microbial biomass
(d) Suppresses denitrification

(ix) Increase in microporosity with the increase in bulk density increases soil water retention

(a) Above field capacity
(b) At field capacity
(c) Below field capacity
(d) None of the above

(x) Time lag between temperature peaks on the soil surface and deeper depths is due to

(a) High volumetric heat capacity of soil
(b) High thermal conductivity of soil
(c) Low thermal conductivity of soil
(d) High thermal diffusivity of soil

(xi) Which one of the following is not a function of soil amendments:

(a) Improvement in soil physical properties
(b) Improvement in soil chemical properties
(c) Improvement in soil biological properties
(d) Improvement in soil mechanical properties

(xii) Polyacrylamide is an example of soil conditioners which behave like

(a) Polycations in soil
(b) Polyanions in soil
(c) Dipole polymers in soil
(d) Emulsions in soil

(xiii) Biochar has a potential of soil carbon sequestration because of

(a) Its highly recalcitrant carbon content
(b) Its easy availability in large amounts
(c) Its easy decomposition in soil
(d) Its strong interaction with soil particles

(xiv) Soil conditioners are primarily used in agriculture to

(a) Decrease soil erosion by water
(b) Improve soil biological properties
(c) Improve soil chemical properties
(d) Improve soil physical properties

(xv) Which one of the following is the most important factor affecting soil physical environment:

(a) Tillage
(b) Mulching
(c) Cropping system
(d) Nutrient management

[**Key:** i. d, ii. c, iii. c, iv. b, v. a, vi. a, vii. b, viii. b, ix. c, x. c, xi. d, xii. c, xiii. a, xiv. d, xv. a]

References

Abid M, Lal R (2008) Tillage and drainage impact on soil quality. I. aggregate stability, carbon and nitrogen pools. Soil Tillage Res 100:89–98

Alkhasha A, Al-Omran A, Aly A (2018) Effects of biochar and synthetic polymer on the hydro-physical properties of sandy soils. Sustainability 10:4642. https://doi.org/10.3390/su10124642

Allen VG, Batello C, Berretta EJ, Hodgson J, Kothmann M, Li X, McIvor J, Milne J, Morris C, Peeters A, Sanderson M (2011) An international terminology for grazing lands and grazing animals. Grass Forage Sci 66:2–28

Ba LT, Van KL, Elsacker SV, Cornelis WM (2016) Effect of cropping system on physical properties of clay soil under intensive rice cultivation. Land Degrad Dev 27:973–982

Bajpai RK, Tripathi RP (2000) Evaluation of non-puddling under shallow water tables and alternative tillage methods on soil and crop parameters in a rice wheat system in Uttar Pradesh. Soil Tillage Res 55:99–106

Blanco-Canqui H, Ruis SJ (2020) Cover crop impacts on soil physical properties: a review. Soil Sci Soc Am J 84(5):1527–1575

Bodman GB, Rubin J (1948) Soil puddling. Soil Sci Soc Am Proc 13:27–36

Brar BS, Singh J, Singh G, Kaur G (2015) Effects of long term application of inorganic and organic fertilizers on soil organic carbon and physical properties in maize–wheat rotation. Agron 5:220–238

Cercioglu M (2017) The role of organic soil amendments on soil physical properties and yield of maize (*Zea mays L.*). Commun Soil Sci Plant Anal 48(6):683–691

Chaudhary NT, Ghildyal BP (1969) Aggregate stability of puddle soil during rice growth. J Indian Soc Soil Sci 17:261–265

Chaudhary NT, Sandhu BS (1982) Soil temperature and plant growth. In: Review of soil research in India, part I. 12th int cong soil sci, New Delhi, India, pp. 48–59

CTIC (Conservation Technology Information Centre) (1999) What's conservation tillage? Lafayette, In: CTIC, Core 4 Program

Deboodt M (1972) Improvement of soil structure by chemical means. In: Hillel D (ed) Optimizing the soil physical environment toward greater crop yields. Academic Press, New York and London, pp 43–55

Dhaliwal SS, Naresh RK, Mandal A, Wakia MK (2019) Effect of manures and fertilizers on soil physical properties, build-up of macro and micronutrients and uptake in soil under different cropping systems: a review. J Plant Nutr 42(3):1–28

Donovan M, Monaghan R (2021) Impacts of grazing on ground cover, soil physical properties and soil loss via surface erosion: a novel geospatial modelling approach. J Environ Manage 287:112206

Douglas JT, Crawford CE (1993) The responses of a ryegrass sward to wheel traffic and applied nitrogen. Grass Forage Sci 48:91–100

Ekeberg E, Riley HCF (1997) Tillage intensity effects on soil properties and crop yields in a long-term trial on morainic loam soil in Southeast Norway. Soil Tillage Res 42:277–293

Ghildyal BP (1982) Nature, physical properties and management of submerged rice soils. In Vertisols and rice soils of the tropics, symposia paper-II. 12 Th int Cong soil sci. Indian soc soil sci, New Delhi, India, pp 121–142

Greene R, Kinnell P, Wood J (1994) Role of plant cover and stock trampling on runoff and soil-erosion from semi-arid wooded rangelands. Soil Res 32:953

Gupta RK, Jaggi IK (1979) Soil physical conditions and paddy yield as influenced by depth of puddling. J Agron Crop Sci 148:329–336

Hayat R, Ali S (2004) Water absorption by synthetic polymer (aquasorb) and its effect on soil properties and tomato yield. Int J Agri Biol 6(6):998–1002

Irmak S, Sharma V, Mohammed AT, Djaman K (2018) Impacts of cover crops on soil physical properties: field capacity, permanent wilting point, soil-water holding capacity, bulk density, hydraulic conductivity, and infiltration. Trans ASABE 61(4):1307–1132

Jamison VC (1953) Changes in air-water relationships due to structural improvement of soils. Soil Sci 76:143–151

Jeffery S, Verheijen FGA, van der Velde M, Bastos AC (2011) A quantitative review of the effects of biochar applications to soils on crop productivity using meta-analysis. Agric Ecosyst Environ 144:175–187

Kalita J, Ahmed P, Baruah N (2020) Puddling and its effect on soilphysical properties and growth of rice and post rice crops: a review. J Pharmacogn Phytochem 9(4):503–510

Lehmann J, Joseph S (2009) An introduction. In: Lehmann J, Joseph S (eds) Biochar for environmental management. Earthscan, London, pp 1–9

Liu E, Yan C, He W, Bing SH, Ding L, Liu Q, Liu S, Fan T (2010) Long-term effect of chemical fertilizer, straw, and manure on soil chemical and biological properties in Northwest China. Geoderma 158:173–180

Liu Z, Cao S, Sun Z, Wang H, Qu S, Lei N, He J, Dong Q (2021) Tillage effects on soil properties and crop yield after land reclamation. Sci Rep 11:4611. https://doi.org/10.1038/s41598-021-84191-z

Majumdar SP (1994) Soil physical constraints and their management in Rajasthan. Tech. Bull. Rajasthan Agri Univ, Jobner Campus, Rajasthan

Majumdar SP, Meena RL, Baghel GDS (2000) Effect of level of compaction and potassium on yield and quality of tomato and chilli crops grown on highly permeable soils. J Indian Soc Soil Sci 48:215–220

Manna M, Swarup A, Wanjari R, Ravankar H, Mishra B, Saha M, Singh Y, Sahi D, Sarap P (2005) Long-term effect of fertilizer and manure application on soil organic carbon storage, soil quality and yield sustainability under sub-humid and semi-arid tropical India. Field Crop Res 93:264–280

Mukherjee A, Lal R (2016) Biochar and soil quality. CRC Press, Boca Raton

Nandapure SP, Sonune BA, Gabhane VV, Katkar R (2011) Long term effects of integrated nutrient management on soil physical properties and crop productivity in sorghum-wheat cropping sequence in a vertisol. Indian J Agric Res 45(4):336–340

Ordoñez-Morales KD, Cadena-Zapata M, Zermeño-González A, Campos-Magaña S (2019) Effect of tillage systems on physical properties of a clay loam soil under oats. Agriculture 9(3):62–76

Peterson GA, Westfall DG, Sherrod LA, Shaver TM (2002) Impact of intensive cropping systems on physical properties of surface soils. In Schlegel AJ (ed) Proc Great Plains Soil Fert Conf, Denver, Colorado, pp 241–247

Pikul JL Jr, Schwartz RC, Benjamin JG, Baumhardt RL, Merril S (2006) Cropping system influences on soil physical properties in the Great Plains. Renewable Agriculture and Food Systems 21(1):15–25

Saha R, Mishra VK, Majumdar B, Laxminarayana K, Ghish PK (2010) Effect of integrated nutrient management on soil physical properties and crop productivity under a maize (Zea mays)–mustard (Brassica campestris) cropping sequence in acidic soils of north-East India. Commun Soil Sci Plant Anal 41:2187–2200

Sharma PK (1988a) Interaction of $FeSO_4$ with polyacrylamide and urea formaldehyde in aggregating light textured soils. Soil Sci 146:185–191

Sharma PK (1988b) Effect of urea formaldehyde on water and air transmission properties of a loamy sand soil. Himachal J Agri Res 14:109–111

Sharma PK (2017) Introduction to soil physics. Westville Publ. House, New Delhi, India, p 512

Sharma PK, De Datta SK (1986) Physical properties and processes of puddled rice soils. Adv Soil Sci 5:139–178

Sharma PK, De Datta SK (1988) Tillage effects on soil physical properties and wetland rice yields. Agron J 80(1):34–39

Sharma PK, De Datta SK (1991) Thermal characterization of a tropical rice soil in relation to puddling, floodwater depth and percolation rate. Soil Tech 4:167–175

Sharma PK, Ingram KT, Harnpichitvitaya D (1995) Subsoil compaction to improve water use efficiency and yields of rainfed lowland rice in coarse-textured soils. Soil Tillage Res 36:33–44

Singh M, Reddy SR, Singh VP, Rupa TR (2007) 2007. Phosphorus availability to rice (*Oryza sativa* L.)–wheat (*Triticum aestivum* L.) in a vertisol after eight years of inorganic and organic fertilizer additions. Bioresour Technol 98:1474–1481

Silburn DM, Carroll C, Ciesiolka CAA, deVoil RC, Burger P (2011) Hillslope runoff and erosion on duplex soils in grazing lands in semi-arid Central Queensland. I. Influences of cover, slope, and soil. Soil Res 49:105

Sollenberger LE, Aiken GE, Wallau MO (2020) Managing grazing in forage livestock systems. In: Management strategies for sustainable cattle production in Southern Pastures, Elsevier Inc. pp 77–100

Steinberger Y, Sarig S, Nadler A, Barnes G (1993) The effect of synthetic soil conditioners on microbial biomass. Arid Soil Res Rehab 7(3):303–306

Sun Q, Meng J, Lan Y, Shi G, Yang X, Cao D, Chen W, Han X (2021) Long-term effects of biochar amendment on soil aggregate stability and biological binding agents in brown earth. Catena 205:105460. https://doi.org/10.1016/j.catena.2021.105460

Teague WR, Apfelbaum S, Lal R, Kreuter UP, Rowntree J, Davies CA, Conser R, Rasmussen M, Hatfield J, Wang T, Wang F (2016) The role of ruminants in reducing agriculture's carbon footprint in North America. J Soil Water Conserv 71(2):156–164

Unger PW, Jones OR (1998) Long term tillage and cropping system affect bulk density and penetration resistance of soil cropped to dryland wheat and grain sorghum. Soil Tillage Res 45:39–57

Unger PW, Eck HV, Musick JT (1981) Alleviating plant water stress. In: Arkin GF, Taylor HM (eds) Modifying the root environment to reduce crop stress. Am Soc Agr Eng, St Joseph, MI, pp 61–96

Verma S (2006) Physical characterization of an acid Alfisol under maize-wheat cropping as influenced by long-term use of chemical fertilizers and soil amendments. PhD thesis submitted to CSK Himachal Pradesh Agricultural University, Palampur (HP), India

Widowati W, Sutoyo, Karamina H, Fikrinda W (2020) Soil amendment impact to soil organic matter and physical properties on the three soil types after second corn cultivation. AIMS Agric Food 5(1):150–168

Wright AL, Hons FM (2005) Tillage impacts on soil aggregation and carbon and nitrogen sequestration under wheat cropping sequences. Soil Tillage Res 84(1):67–75

Wu Q, Xie HL (2017) A review and implication of land fallow systems research. J Resour Ecol 8:223–231

Zein El-Abdeen HA (2018) Interference between organic soil conditioners mixed with synthetic soil conditioners to improve sandy soil productivity. J Soil Sci Agric Eng Mansoura Univ 9(12):723–734

Zhang S, Li Q, Zhang XP, Wei K (2012) Effects of conservation tillage on soil aggregation and aggregate binding agents in black soil of Northeast China. Soil Tillage Res 124:196–202

Zhou M, Liu C, Wang J, Meng Q, Yuan Y, Ma X, Liu X, Zhu Y, Ding G, Zhang J, Zeng X, Du W (2020) Soil aggregates stability and storage of soil organic carbon respond to cropping systems on black soils of Northeast China. Sci Rep 10:265. https://doi.org/10.1038/s41598-019-57193-1

MIX
Papier aus verantwortungsvollen Quellen
Paper from responsible sources
FSC® C105338

If you have any concerns about our products,
you can contact us on
ProductSafety@springernature.com

In case Publisher is established outside the EU,
the EU authorized representative is:
Springer Nature Customer Service Center GmbH
Europaplatz 3, 69115 Heidelberg, Germany

Printed by Libri Plureos GmbH
in Hamburg, Germany